Controlled Release Systems: Fabrication Technology

Volume I

Editor

Dean Hsieh, Ph.D
President
Conrex Pharmaceutical Corp.
Brandamore, Pennsylvania

CRC Press, Inc.
Boca Raton, Florida

Library of Congress Cataloging-in-Publication Data

Controlled release systems.

   Bibliography: p.
   Includes indexes.
   1. Drugs--Controlled release. 2. Controlled release
preparations. I. Hsieh, Dean, 1948-
RS201.C64C665   1988   615'.19   87-21785
ISBN 0-8493-6013-7 (v. 1)
ISBN 0-8493-6014-5 (v. 2)

This book represents information obtained from authentic and highly regarded sources. Reprinted material is quoted with permission, and sources are indicated. A wide variety of references are listed. Every reasonable effort has been made to give reliable data and information, but the author and the publisher cannot assume responsibility for the validity of all materials or for the consequences of their use.

Direct all inquiries to CRC Press, Inc., 2000 Corporate Blvd., N.W., Boca Raton, Florida, 33431.

International Standard Book Number 0-8493-6013-7 (volume I)
International Standard Book Number 0-8493-6014-5 (volume II)
Library of Congress Card Number 87-21785
Printed in the United States

FOREWORD

Controlled release technology is a rapidly emerging field. Two decades ago, formulations based on this technology were scarcely in existence. Today, the number of controlled release products is large and growing rapidly. With the advent of biotechnological advances and genetic engineering, the development of new and more complex drugs is imminent. These developments have necessitated more than ever before the creation of effective delivery systems which can protect these precious molecules from destruction, yet continuously deliver them to the body safely. Central to the successful development of any controlled release system is the fabrication of formulation procedure — it must be safe, reproducible, not damaging to the drug, and amenable to scale-up. The choices as to the types of controlled release systems one might use and the different ways these systems can be fabricated are significant. Such choices can only be made once an understanding of the principles underlying these systems and fabrication procedures is in hand.

While many books have been written on controlled release, this book stands alone in its effort to bring together an understanding of the ways in which controlled release systems can be fabricated. Dr. Hsieh is certainly well qualified to edit and put together such a book, as he is a skilled formulator of controlled release systems and has studied them extensively both as a postdoctoral scientist in our laboratory at MIT, as a Professor at Rutgers, and at his company, Conrex Pharmaceutical Corporation. This book brings something new to readers in the exciting field of controlled release technology/fabrication technology, an area that should be of practical laboratory value in the design of these important systems.

Robert Langer

PREFACE

Controlled release technology has well deserved its recent growth in popularity and widespread acclaim. Its advantages have been recognized and utilized not only by the pharmaceutical industry, but by several other industries as well. Advances in controlled release research enable innovative refinements in many currently manufactured household products. Employment of these refinements helps the sponsors of such research to achieve a competitive edge. However, before controlled release products can be manufactured, further research and development must take place. Fabrication technology refers to the methods by which controlled release products are manufactured. Throughout the development of controlled release devices, fabrication technologies have played a key role in the process of innovation. The purpose of these volumes is to compile and generalize principles of fabrication methods which have been previously published. These volumes thus provide a framework for the study of fabrication technology. It is the editor's hope that they will form the basis for future innovation.

The first volume is concerned with fabrication procedures for currently marketed products or mature technologies. The second volume is concerned with fabrication procedures in various stages of development. Some of the technologies described in the second volume may be mature, yet may belong to a class of products wherein the majority are still under development. Volume I, Chapter 1 is concerned with this classification and explores it in detail. Briefly, there are three stages in the development of controlled release technology:

1. Encapsulation technologies are covered in Chapters 2 through 7. These fabrication procedures have matured, now comprising, for instance, coacervation, film coating, and mechanical blending.
2. Transdermals and other advanced drug delivery systems are covered in Chapter 8 and Volume II, Chapters 1 through 7. These include multiple lamination for transdermal patches, injection molding, extrusion, gelation, multiple emulsion, and other methods for fabricating bioerodible and hydrogel drug delivery systems.
3. Selective drug targeting systems are surveyed in Chapter 8. These systems represent the latest and most promising stage in controlled release technologies. They include monoclonal antibodies, liposome delivery systems, dextran and magnetic microspheres, and polymeric delivery systems.

All three stages in controlled release technology must utilize various sterilization procedures according to individual fabrication processes. Chapter 9 provides an overview of sterilization procedures for controlled release products.

Much of the knowledge contained within these volumes will prove valuable to the scientific community. For this reason, it has been difficult for contributors to disclose proprietary information. The organizations and individuals who have offered the results of costly and painstaking research deserve greater rewards than I can offer. Nevertheless, I take this opportunity to thank all the contributors for the time and effort they have devoted to this project. I also thank those friends whose guidance directed me to these outstanding individuals. These volumes are truly the product of a concerted effort by a talented and enthusiastic group of colleagues. Those with expertise and a willingness to share it have made this book possible to complete. Even those authors whose companies vetoed their contributing chapters added stimulus to the project with their initial enthusiasm.

In addition to the major contributors to these volumes, there have been several individuals whose continuous support has sustained me through the project. I thank my wife, Mrs. Phyllis Hsieh, for her unqualified patience, endurance, and encouragement. I also thank J. C. Lorber for the capable and prudent input as well as the cooperative effort necessary to make constant and steady progress. I would also like to acknowledge the assistance of my colleagues at the Rutgers College of Pharmacy: Drs. Y. Chien, K. Tojo, and C. Liu, among others. Their criticism concerning the pursuit of this project, both positive and negative, has been appreciated. This appreciation extends to several graduate students at Rutgers, including P. Mason, C. C. Chiang, E. Tan, and R. Bogner, who helped to proofread the communications involved in the project. Furthermore, I must thank the editorial staff at CRC Press, Inc., for the experienced coordination of communication which led to final agreements with the contributing authors. Finally, my thanks to CRC Press for giving me this opportunity to pursue this most significant and greatly rewarding project.

Dean Hsieh

EDITOR

Dr. Dean Hsieh is the founder and president of Conrex Pharmaceutical Corporation, Brandamore, Pennsylvania since 1985. He obtained his Ph.D. ('78) and M.S. ('74) degrees from M.I.T., followed by postdoctoral training with Professor Langer at the Boston Children's Hospital, associated with the Harvard Medical School. In January, 1981, he was promoted to Instructor at Harvard Medical School. In October 1982, he became an Assistant Professor at the College of Pharmacy, Rutgers — The State University of New Jersey. He has authored and co-authored more than 60 papers and abstracts in the area of drug delivery systems and controlled release technologies. He is also the holder of several patents and pending patents. Recently, his efforts have focused on the research and development of proprietary permeation enhancers, leading to the commercialization of this technology.

CONTRIBUTORS

Jones W. Fong, Ph.D.
Associate Fellow
Department of Pharmaceutical
  Development
Sandoz Research Institute
East Hanover, New Jersey

Eric Forster
Technical Service
Manesty Machines, Limited
Liverpool, England

Robert P. Giannini, Jr., Ph.D.
Director of Pharmaceutics
hiMEDICS
Hollywood, Florida

Harlan S. Hall
President
Coating Place, Inc.
Verona, Wisconsin

Dean S. T. Hsieh, Ph.D.
President
Conrex Pharmaceutical Corporation
Brandamore, Pennsylvania

Takafumi Ishizaka, D. Pharm.
Assistant
Faculty of Pharmaceutical Science
Science University of Tokyo
Tokyo, Japan

Masumi Koishi, D.Sc.
Professor
Faculty of Industrial Science and
  Technology
Science University of Tokyo
Oshamanbe, Hokkaido, Japan

Aracelis Maria Ortega
FARM S.A.
Caracas, Venezuela

Pramod P. Sarpotdar, Ph.D.
Eastman Kodak
Rochester, New York

Carl R. Steuernagel
Product Manager
Food and Pharmaceutical Division
FMC Corporation
Philadelphia, Pennsylvania

Harry S. Thacker
Technical Director
Manesty Machines, Limited
Liverpool, England

Jesse Wallace
Consultant
Food and Pharmaceutical Division
FMC Corporation
Philadelphia, Pennsylvania

TABLE OF CONTENTS

Volume I

Volume II

Chapter 1

CONTROLLED RELEASE SYSTEMS: PAST, PRESENT, AND FUTURE

Dean S. T. Hsieh

TABLE OF CONTENTS

I. DECLINE IN NEW DRUG DEVELOPMENT

Though significant progress in new drug development would not escape the attention of the general public, current public awareness of the state of such research is minimal at best. The absence of recent spectacular successes and dramatic discoveries has caused public interest in the pharmaceutical industry to wane. Such absence may be attributed to a wide variety of causes. These include the following:

1.   The cost in time and money of new drug development has escalated over the past two decades.
2.   Increased regulatory control has slowed the drug development process in the U.S.
3.   New drug development requires the cooperation of a wide variety of practitioners: chemists, biochemists, pharmacologists, and clinicians. All these scientists realize that their chances of producing a new drug are statistically very low.[1]

The development of new drugs is indeed a slow and difficult process. After its original discovery, a drug must pass a biological screening and animal toxicity and stability studies. Only then may an investigational new drug (IND) application be prepared and filed with the Federal Food and Drug Administration (FDA). A drug must then pass more thorough toxicity studies before it may be used in human clinical studies. Human clinical studies may be delineated in three stages:

1.   Phase I — Toxicity, metabolism absorption, and safe dosage ranges (among other things) are evaluated.
2.   Phase II — Early trials are conducted to determine optimum doses and optimum formulation.
3.   Phase III — A wide range of clinical trials are conducted to assess safety and effectiveness for treatment of a given disease.

Once a drug has successfully completed Phase III, a new drug application (NDA) may be prepared and filed with the FDA. NDA approval results in marketing and distribution.

The mean time between filing a new chemical entity (NCE) IND application and receiving NDA approval from the FDA has steadily increased since the addition of the Kefauver-Harris Amendments to the Food, Drug, and Cosmetic Act in 1962.[3] Moreover, only 2 of the 232 NCE INDs accepted for human clinical study in 1968 had achieved NDA approval by 1974.[4] Many studies point to increased federal regulation as the cause of the long waiting period.[5-7] This finding is supported by the existence of a statistically significant difference in time between the introduction of a beneficial new drug in foreign countries and the introduction of the same drug in the U.S. More often than not, the U.S. finds itself on the slow end of a "drug lag". A study by Wardell and Lasagna reveals that Britain had double the U.S. lead time in the introduction of beneficial new drugs from 1961 through 1971.[8] Furthermore, at the end of 1971, there were four times as many single chemical entities exclusively available in Britain than there were available in the U.S.

That the development of a new drug is an extremely long and difficult process is evidenced by the fact that in 1970 the U.S. pharmaceutical industry prepared, extracted, or isolated for medical research purposes 126,060 new drug substances; 703,900 substances were submitted for pharmacological study, yet only 1013 were found useful and selected for clinical testing. Only 16 were ultimately selected for NDA approval.[9]

The substantial regulatory requirements which must be met prior to the introduction of a new drug in the U.S. have made it very difficult for pharmaceutical companies to earn a return on their research and development investments. Clymer claims that a return may not appear for 19 years after the initial marketing.[10]

The restricted atmosphere typical to a new drug development in the U.S. has prompted many pharmaceutical companies to move their investments abroad. U.S. companies are attracted by a foreign market which is, collectively, three times the size of the U.S. market. Though regulatory restrictions are on the rise in foreign countries as well, those of the FDA have always been much more stringent.[11] Despite protests that the U.S. regulatory system is hindering the progress of science, there is no indication that a reevaluation of the present system is imminent.

The cost involved in developing new drugs is keeping pace with the increase in regulation. In 1960, the pharmaceutical industry invested less than $100 million and was able to generate 50 new drug entities. In 1965, a $350 million investment produced only 25 new drug entities. By 1975, a $1.028 billion investment resulted in just 15 new drug entities. Research and development effectiveness in the drug industry now costs much more time as well as money. In 1960, it took an average of 2 years to develop a new single chemical entity; in 1975, it took between 10 and 15 years.[12]

The 1980s have brought radical changes to the pharmaceutical industry. Ever-increasing drug development costs have persuaded the U.S. Department of Justice to condone joint research ventures. In a 1981 publication, "Anti-Trust Guide Concerning Joint Research Ventures," the Department asserts that anti-trust laws prevent few, if any, cooperative research ventures among firms in the same or different industries.[13] These guidelines enable small or medium-sized pharmaceutical companies to engage in drug research — an option that had previously been ruled out by escalating costs.

Another option for companies wishing to engage in profitable research is evidenced by recent advances in the area of medical devices.[14] This field has long been populated by quacks promoting useless contraptions as treasures of doubtful therapeutic value. However, modern technology has enabled researchers to merit FDA approval for tooth sealants, hypothermic devices, and nuclear magnetic resonance (NMR). Because research in this area is less costly and potentially more rewarding than development of a new drug entity, the R & D dollars of the pharmaceutical industry are rapidly shifting toward drug delivery systems and away from traditional drug research.[15] The skyrocketing price on new drug development has also led the pharmaceutical industry to seek methods of improving the administration of existing drugs. This research has been stimulated by the expiration of patent rights on the largest selling drugs in today's market.[16] The 25 largest-selling post-1962 prescription drugs had a collective estimated 1983 sales figure of $1.7 billion. These 25 individual markets include Inderal® (1983 sales estimate: $280 million), Dyazide® (1983 sales estimate: $250 million), and Aldomet® (1983 sales estimate: $175 million)[17] (see Table 1). Both the companies who hold patents on these drugs and their competitors are endeavoring to develop new timed release or sustained release methods of administration. Patentable success in the development of novel timed release drug administration systems may lead to unprecedented rapid gains in the distribution market for an individual drug or group of drugs.

Another factor which makes device technology particularly alluring is the recent impact of the FDA on the nonprescription drug market.[19] Nonprescription drugs available today are safe and effective. Much effort is being made to educate today's health-conscious public on the proper use of these drugs. Major technological breakthroughs in the form of drug administration devices would receive immediate recognition and deliver immediate rewards to companies willing to engage in such research.

II. ADVANTAGES OF DRUG DELIVERY RESEARCH

Controlled release administration denotes a method by which one preparation of a drug

Table 1
THE 25 LARGEST-SELLING PRESCRIPTION DRUGS
CONSIDERED MOST SUSCEPTIBLE TO GENERIC COMPETITION[18]

Brand name	Generic name	Brand firm	Estimated 1983 sales (millions)
Inderal®	Propranolol	Ayerst	280
Dyazide®	Triamterene + hydrochlorothiazide	SmithKline	250
Aldomet®	Methyldopa	Merck	175
Diabinese®	Chlorpropamide	Pfizer	120
Lasix®	Furosemide	Hoechst	100
Darvocet N®	Propoxyphene + napsylate acetaminophen	Lilly	95
Indocin®	Indomethacin	Merck	90
Vibramycin®	Doxycycline	Wyeth	75
Ativan®	Lorazepam	Wyeth	75
Aldoril®	Methyldopa + hydrochlorothiazide	Merck	75
Tolinase®	Tolazamide	Upjohn	70
Theo-Dur®	Theophylline	Key	60
Procan SR®	Procainamide	Parke-Davis	60
Catapres®	Clonidine	Boehringer-Ingelheim	55
Triavil®	Perphenazine + amitriptyline	Merck	55
Dalmane®	Flurazepam	Roche	50
Hygroton®	Chlorthalidone	USV	42
Ortho-Novum® 35-28	Norethindrone combination	Ortho	40
Tegretol®	Carbamazepine	Geigy	40
Ortho-Novum® 50-21	Norethindrone combination	Ortho	37
Nolvadex®	Tamoxifen	Stuart	35
Zyloprim®	Allopurinol	Burroughs Wellcome	35
Ortho-Novum® 35-21	Norethindrone combination	Ortho	35
Synthroid®	Levothyroxine	Flint	35
Nitro-Bid®	Nitroglycerine	Marion	30

will accomplish its desired therapeutic effect with more consistency and longer duration than the usual multiple doses of the same drug. The effectiveness of a drug administration method is measured by the level of activity indicative of its desired pharmacological response. In some cases, this may be measured by the concentration of the drug in the bloodstream.[20] As a usual single dose is ingested, drug activity increases steadily. It may approach toxic levels as the maximum concentration is achieved. Once bodily absorption has countered this threat, drug activity is steadily reduced to a level which is ineffective in treating the patient. The second dose sends the drug concentration climbing once again and the hill-and-valley pharmacological phenomenon is repeated (see Figure 1).

The aim of controlled release technology is to eliminate this cyclic aspect of modern drug administration. A controlled release capsule releases a quantity of the drug which has been determined to achieve the optimal concentration. It supplements the initial burst at a rate equal to that of bodily absorption. Only one oral dose is required and toxic effects and periods of ineffective treatment are precluded.

The advantages which stem from a prolonged period of therapeutically effective drug activity are obvious. Reasons for favoring one oral dose over many include: greater patient

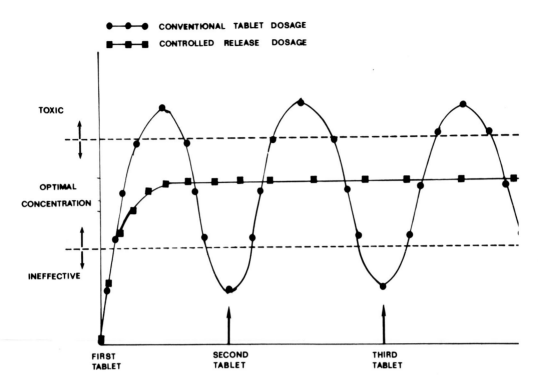

FIGURE 1.   Hill-and-valley pharmacological phenomena resulting from conventional tablet dosage (●). Steady-state blood concentration resulting from controlled release dosage (■).

compliance, saving the time of nurses and pharmacists, and conserving hospital storage space.[21] In addition, therapy would be enormously improved if drugs could somehow be equipped with target-selective homing ligands. The so-called "magic bullet" concept has been the dream of research pharmacologists since its conception by Paul Ehrlich.[22] Target selectivity in drug administration has traditionally been achieved by either differential sensitivity or differential accessibility.[23] The property of differential sensitivity is ascribed to a drug which spreads throughout the body but acts only on the afflicted area. That of differential accessibility is ascribed to a drug which may act on the entire system, but is applied solely to the afflicted area. Diseases whose structure cannot be distinguished from human cellular structure must be treated by differential sensitivity. It has been the task of pharmaceutical research to isolate drugs which are toxic to parasitic and destructive cells, but neutral or beneficial to humans. However, methods have recently been devised to differentiate such cells on the surface.[24] It is now theoretically possible to use differential accessibility as a tool in combating previously undistinguishable cells, for example, those of viruses and cancers. Modern pharmacology envisions selective devices as drug carriers. Such devices would render traditional drug administration methods both inefficient and wasteful.

III. TECHNOLOGICAL BREAKTHROUGHS

A. Encapsulation: Microencapsulation and Macroencapsulation

Throughout the 1960s and the 1970s, research and development activity on microencapsulation techniques and applications flourished (see Figure 2). Dozens of patents were issued for microencapsulation innovations ranging from rust inhibitors to silver capsules for dental surgery. Researchers were quick to report their findings. Many books were published as well as innumerable articles. The result of this rapid progress is evident in the percentage

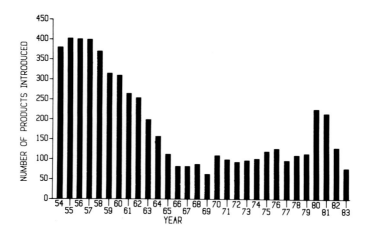

FIGURE 2.   Number of new drugs introduced form 1954 to 1983.

of controlled release drugs which rely on microencapsulation for their sustained release characteristics (see Figures 2 and 3). More than half (65%) of controlled release drugs now belong to this category (see Table 2). Moreover, applications for microencapsulation technology are not restricted to drug delivery systems. Much of the microencapsulation research in the 1970s concentrated on biomedical applications. It was discovered that microencapsulated living cells and tissues may continue to grow and perform normal functions.

Developments in microencapsulation technology can be viewed as the natural consequence of two major technological breakthroughs. In 1949, the Wisconsin Alumni Research Foundation submitted a patent for the widely adopted Wurster process. Dale Wurster, then of the University of Wisconsin, had utilized a fluidizing bed and a drying drum to encapsulate fine solid particles suspended in midair.[25] This discovery prompted numerous research projects in both academia and industry. Just 4 years later, NCR Corporation submitted two patent applications for the coacervation method which was developed by B. K. Green at their laboratory. Green invented a new type of carbon paper, one which produced copies by coating the back of the original with an acidic dye and the surface of the duplicate with an encapsulated colorless dye base.[26] Though not optimally efficient, this invention was significant because liquids had not been encapsulated previous to Green's efforts. The wide variety of potential applications for microencapsulation technology made it the subject of intensive and competitive research for the decade following Wurster's discovery. Between 1956 and 1966, over 50 patents were filed for innovative microencapsulation techniques.

Most of the processes discovered at that time can be classified as physical microencapsulation processes or mechanical methods. The Wurster process is a mechanical method. Other mechanical methods include pan coating, the gravity-flow method, and centrifugal methods. The pan-coating method produces relatively large capsules (1 mm in diameter). The microencapsulated substance is coated onto spherical substances such as nonpareil sugar seeds and the coating is applied in the form of a solution or an atomizing spray.[27] The final product is not usually optimally uniform. The gravity-flow method is a relatively slow process wherein a liquid coating is allowed to form a membrane on the bottom of an open vertical cylinder. The material to be coated is introduced at the top of the cylinder. It falls into the membrane, causing the membrane to break from the cylinder. Surface tension causes the capsule to assume a spherical shape as it falls into a hardening medium.[28] Centrifugal processes also utilize fluid membranes across orifices; however, they are formed on the sides of a rapidly rotating cylinder while the material to be encapsulated is at its core. The material is flung through the membrane into the hardening medium.[29]

More recently, microencapsulation research has concentrated on the discovery and de-

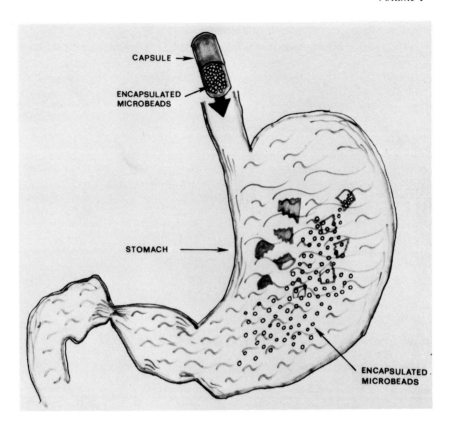

FIGURE 3.   Microencapsulation for sustained release preparations.

velopment of semipermeable microcapsules, particularly on semipermeable encapsulation methods which do not require processing with either heat or organic solvents.[30] These restrictions yield a technique capable of encapsulating living cells. Encapsulation of living cells is desirable for several reasons, including the following:[31]

1.   They may be treated as individual microculture flasks and are easier to handle and manipulate.
2.   Bacterial and viral cells cannot cross the semipermeable shell; both in vivo and in vitro sterile conditions are ensured.
3.   Cell permeability may be adjusted as needed for a particular application.

Potential applications for microencapsulated living cells are also numerous. Encapsulated microorganisms may produce antibiotics, amino acids, and other useful chemicals; this process is called fermentation. Microencapsulated insulin-producing islets have produced insulin for up to 8 weeks — 5 weeks longer than nonencapsulated islets.[32] Encapsulated tumor cells may continuously release antigens which trigger antibody production in animals. In the words of cell-encapsulation pioneer T. M. S. Chang, "The potential of artificial cells in biomedical research is limited only by one's imagination."[33]

Perhaps because of the declining profitability of new drug development, the 1970s saw a great increase in the development of microencapsulation techniques. The market for such products grew as researchers sought new ways to exploit their advantages and practitioners became aware of their benefits. By the 1980s, this market had matured; microencapsulation technology had fulfilled its promise.

Table 2
CONTROLLED RELEASE PRODUCTS BASED ON ENCAPSULATION

Therapeutic uses	Drug	Trade name	Company
Analgesic	Hydrocodone bitartrate	Bancap HC® Capsules	O'Neal, Jones & Feldman Pharmaceutical
	Acetaminophen		
	Aspirin	Arthritis Bayer® Timed-Release	Sterling Drug
Anorexiant	Methamphetamine hydrochloride	Desoxyn®	Abbott Laboratories
	Phentermine hydrochloride	Fastin® Capsules	Beecham Laboratories
	Phentermine resin	Ionamin®	Pennwalt
	Diethylpropion hydrochloride	Tepanil® Ten-tab®	Riker (3M Company)
	Phenylpropanolamine hydrochloride	Control Capsules	Thompson Medical Company, Inc.
		Dexatrim® Capsules	Thompson Medical Company, Inc.
	Caffeine		
Antiallergenic	Chlorpheniramine maleate	Dallergy® SR Capsules	Laser, Inc.
	Phenylephrine hydrochloride		
	Methscopolamine nitrate		
Antiangina	Isosorbide dinitrate	Iso-Bid® Capsules	Geriatric Pharmaceutical
Antianxiety	Meprobamate	Meprospan®	Wallace Labs (Carter-Wallace)
Antiarrhythmic	Quinidine gluconate	Duraquin®	Parke Davis (Warner Lambert)
Antiasthmatic	Anhydrous theophylline	Theo-Dur®	Key Pharmaceuticals
	Pseudoephedrine hydrochloride	Respaire®-SR Capsules	Laser, Inc.
	Theophylline anhydrous	Theospan®-SR Capsules	Laser, Inc.
Antidepressant	Lithium carbonate	Lithobid®	Ciba-Geigy
Antihistamine	Chlorpheniramine maleate	Isoclor® Timesule® Capsules	Fisons
	Pseudoephedrine hydrochloride		
	Phenylpropanolamine hydrochloride	Naldecon®	Bristol-Myers
	Phenylephrine hydrochloride		
	Phenyloxamine citrate		
	Chlorpheniramine maleate		
	Tripelennamine hydrochloride	PBZ-SR®	Ciba-Geigy
	Chlorpheniramine maleate	Sinovan Timed®	Drug Industries Company
	Phenylephrine hydrochloride		
	Methscopolamine nitrate		
	Guaifenesin	Congess Sr. Capsules	Fleming and Company
	Phenylephrine hydrochloride	Extendryl	Fleming and Company
	Pseudoephedrine		
	Methscopolamine nitrate		
	Chlorpheniramine maleate	Pseudo-hist	Holloway Pharmaceuticals
	Pseudoephedrine hydrochloride		
	Chlorpheniramine maleate	Histor-D® Timecelles	W. E. Hauck, Inc.

Table 2 (continued)
CONTROLLED RELEASE PRODUCTS BASED ON ENCAPSULATION

Therapeutic uses	Drug	Trade name	Company
	Phenylephrine		
	Methscopolamine nitrate		
	Brompheniramine maleate	Bromfed®	Muro Pharmaceuticals
	Pseudoephedrine hydrochloride		
	Anhydrous theophylline	Theolair®-SR	Riker (3M)
	Brompheniramine maleate	Dimetapp® Elixir	A. H. Robins Company
	Phenylephrine hydrochloride		
	Phenylpropanolamine		
	Pseudoephedrine hydrochloride	Fedahist® Gyrocaps®	William H. Rorer, Inc.
	Chlorpheniramine maleate		
	Theophylline anhydrous	Slo-Phyllin® Gyrocaps®	William H. Rorer, Inc.
	Guaifenesin		
	Carbinoxamine maleate	Rondec-TR®	Ross Laboratories
	Pseudoephedrine hydrochloride		
	Diphenylpraline hydrochloride	Hispril® Spansule®	Smith Kline and French
	Chlorpheniramine maleate		
	Caramiphen edisylate	Tuss-Ornade®	Smith Kline and French
	Phenylpropanolamine hydrochloride		
	Chlorpheniramine	Histaspan-D®	USV (Revlon)
	Phenylephrine hydrochloride		
	Methscopolamine nitrate		
	Chlorpheniramine maleate	T-Dry	T. E. Williams Pharmaceuticals, Inc.
	Phenylpropanolamine hydrochloride		
	Phenylephrine hydrochloride		
Antihypertensive	Propranolol hydrochloride	Inderal®	Ayerst (AHP)
Antispasmodic	Hyoscyamine sulfate	Levsinex® Timecaps®	Kremers-Urban
	Trihexyphenidyl hydrochloride	Artane®	Lederle (Cyanamid)
	Hyoscyamine sulfate	Donnatal	A. H. Robins
	Atropine sulfate	Extentabs®	
	Scopolamine hydrobromide		
	Prochlorperazine maleate	Combid® Spansule®	Smith Kline and French
	Isopropamide iodide		
14-hr aspirin	Aspirin	Verin®	Verex Laboratories, Inc.
Bronchial asthma	Theophylline anhydrous	Bronkodyl® S-R	Winthrop-Breon Laboratories (Sterling)
Bronchodilator	Theophylline anhydrous	Constant-T®	Ciba-Geigy
		Somophyllin®-CRT Capsules	Fisons
		Theobid® Duracap®	Glaxo, Inc.
		Theophyl®-SR	McNeil Pharmaceutical
		Quibron®-T/SR	Mead-Johnson (Bristol-Myers)

Table 2 (continued)
CONTROLLED RELEASE PRODUCTS BASED ON ENCAPSULATION

Therapeutic uses	Drug	Trade name	Company
		Sustaire®	Roerig (Pfizer)
Cardiovascular	Isorbide dinitrate	Isordil®	Ives (AHP)
	Papaverine hydrochloride	Pavabid®	Marion
	Quinidine sulfate	Quinidex Extentabs®	A. H. Robins
	Nitroglycerine	Nitrong®	Wharton Labs, Inc. (U.S. Ethicals)
CNS stimulant	Dextroamphetamine sulfate	Dexedrine® Spansule®	Smith Kline and French
CNS	Perphenazine	Trilafon® Repetabs®	Schering Corporation
Decongestant	Chlorpheniramine maleate d-Pseudoephedrine hydrochloride	Deconamine®-SR	Berlex Labs, Inc.
	Brompheniramine maleate Phenylephrine hydrochloride Phenylpropanolamine hydrochloride	Brocon C.R.	Forest Labs, Inc.
	Pseudoephedrine hydrochloride Guaifenesin	Pseudo-Bid	Frye Pharmaceuticals
Glaucoma	Acetazolamide	Diamox®	Lederle (Cyanamid)
Hormone	Methyltestosterone	Android-5®	Brown Pharmaceutical Company, Inc.
	Methyltestosterone	Oreton® Methyl	Schering Corporation
"Hot flashes"	Phenobarbital Alkaloids of belladonna	Bellergal-S®	Sandoz Pharmaceuticals
Muscle relaxant	Isosorbide dinitrate	Sorbitrate®	Stuart Pharmaceuticals
	Papaverine hydrochloride	Cerespan®	USV (Revlon)
Myasthenia	Pyridostigmine bromide	Mestinon®	Hoffman-La Roche
Nutrition	Potassium chloride	Kaon CL-10®	Adria Labs
	Vitamin B-1, B-6	B-C-Bid Capsules	Geriatric Pharmaceutical Corp.
Geriatric	Niacinamide Calcium pantothenate Vitamin C, B-12		
	Vitamin C	Cevi-Bid®	
	Ferrous fumarate	Ferro-Sequels®	Lederle (Cyanamid)
	Potassium chloride	Klotrix®	Mead-Johnson (Bristol-Myers)
	Potassium chloride	Micro-K Extencaps®	A. H. Robins
	Potassium chloride	Slow-K®	Ciba-Geigy
	Ferrous sulfate	Feosol® Spansule®	Menley (Smith Kline Consumer)
	Chlorpheniramine maleate	Chlorafed Timecelles	W. E. Hauck, Inc.
Stimulant	Caffeine	No Doz®	Bristol-Myers
Vasodilator	Isosorbide dinitrate	Dilatrate®-SR	Reed & Carnrick

Note: The information in this table was compiled from the *Physician's Desk Reference*, 38th ed., Medical Economics, 1984.

Table 3
TRANSDERMALS AND OTHER ADVANCED DRUG DELIVERY SYSTEMS

Therapeutic uses	Drug	Trade name	Company
Transdermals			
Cardiovascular	Nitroglycerine	Nitro-Dur®	Key Pharmaceuticals
		Transderm-Nitro®	CIBA Pharmaceutical Company
	Nitroglycerine	Nitrodisc®	Searle Pharmaceuticals
Motion sickness	Scopolamine	Transderm-Scōp®	CIBA Pharmaceutical Company
Antihypertension	Clonidine	Catapres®ᵃ	Boehringer Ingelheim Ltd.
Antimenopause	Estradiol	ᵃ	ᵃ
Others			
Delivery system		Flexiflo®	Ross Laboratories
Dry eye syndrome	Hydroxypropyl cellulose	Lacrisert Sterile Ophthalmic Insert®	Merck Sharp and Dohme

Note: The information in this table was compiled from the *Physician's Desk Reference*, 38th ed., Medical Economics, 1984.

ᵃ Not found in the *Physician's Desk Reference*, 1984.

B. Transdermals

Research is now concentrating on transdermal delivery systems. It is easy to tell why. One of the first successful uses of transdermal patch technology was a nitroglycerine cardiatric application.[34] It was introduced in 1982 and by 1984 the market for nitropatches had already soared to $200 million. The 1980s is the decade for transdermal patches to emerge from the laboratory and capture their waiting markets (see Table 3).

There exist several types of designs for transdermal patches, but all consist of the same basic components: a backing film, a drug reservoir, a membrane, and a contact adhesive. The patches administer drugs through the skin. They are attached to the skin via the contact adhesive, and the drug is absorbed from the reservoir by the skin and passed through capillaries under the surface of the skin to the bloodstream.

There are a number of reasons that transdermal patches have become the subject of major research efforts. Despite progress made in microencapsulation techniques, any form of drug absorption through the gastrointestinal tract is inherently unreliable. GI tract absorption varies with the amount of food in the stomach, GI motility, the time it takes a capsule to reach its destination, and most notably possible deactivation by first-pass metabolism before reaching systemic circulation.[35] Variable GI tract absorption results in unpredictable drug concentration levels in the bloodstream. This phenomenon calls for a method of introducing drugs directly into the systemic circulation. Cited advantages of transdermal patches include:

1. Minimization of drug exposure by allowing a significant reduction in dosage
2. Provision of long-term therapy from a single dose and thus increase in patient compliance
3. Avoidance of the risks and inconveniences of intravenous therapy
4. Possibility of the use of drugs with short biological half-lives
5. Allowance of immediate termination of drug input by simply removing the patch

There currently exist three types of transdermal patches: reservoir systems, matrix systems, and microsealed systems.[36] Reservoir systems consist of solid particles suspended in a liquid medium which is enclosed by an impermeable membrane laminate. The rate of drug release is controlled by a microporous or nonporous polymeric membrane. In a matrix system, solid drug particles are dispersed in a diffusion-controlled matrix medium formed of gels or polymers. This is enclosed by an impermeable membrane laminate, a fixed portion of which exposes the matrix to permit the continuous release of drug molecules. The matrix system is safer than the reservoir system, for there is no danger of the reservoir bursting and causing a flood of the drug to be released. A microsealed system is a polymer matrix which stores drugs in millions of microscopic liquid compartments. This device can be molded into virtually any shape. Its reliability may be enhanced by covering the exposed surface of the matrix with a permeation-controlling polymeric membrane.

Though the FDA treats all transdermal controlled release medicines as new drugs, subject to the same intensive regulation and standards, the fact that transdermal patches are formulated using existing approved drugs smooths the way for the development process. For example, if the systemic toxicity of the therapeutically active agent is already well-documented, the FDA may waive the requirement for animal toxicity studies and simply require additional clinical testing on humans.[37] Systemic safety issues may be ruled out altogether when transdermal doses place drug concentration in the blood at levels well below those of currently approved labeled dosages. However, additional metabolic studies may be required if there exists a possibility that bypassing the GI tract, the difference in hepatic-portal metabolism, or the role of the kidney in the elimination of the drug from the body significantly alters the effect of the drug in transdermal drug administration.[38] Final approval for transdermal patches is based on three major considerations:

1.   Reproducibility of plasma levels
2.   Defined pharmacokinetic parameters to support drug labeling
3.   Demonstration that the plasma concentration is within established therapeutic limits for a particular drug entity

Reproducible plasma levels are not as easy to achieve as one may expect. There do not exist biological factors which influence the rate of drug absorption from transdermal patches. For a generalized instance, the skin has limitations on the amount of drug which it may absorb. This factor may sometimes be dealt with by increasing the size of the patch. In cases where it is imperative that the delivery system control drug input, treatment is designed assuming the least permeable absorption rate.[40]

Variation among drug absorption rates in people remains a problem, however. Although there seems to be a correlation between the number of cell layers in the stratum corneum and drug permeability, the shape and configuration of those cells presents an equally significant factor in absorption.[41] Moreover, the percutaneous absorption rate is highly dependent on the nature of the drug. Those that are relatively water insoluble may dissolve in the stratum corneum and become an integral component of the membrane.[42] Some drugs have even been detected 4 weeks after exposure to the skin. Moreover, substances have been identified which allow researchers to alter the rate of drug permeation.[43] Dedicated research efforts will doubtless fully explore all the variables involved in percutaneous absorption as they refine transdermal drug delivery systems to meet FDA standards.

C. Other Advanced Drug Delivery Systems

Efforts to exploit existing drugs rather than develop new ones range far beyond the scope of transdermal technology. A selective sample of innovative new drug delivery techniques is given in Table 4, any one of which may serve to capture the market for an individual drug or set of drugs. These technological innovations are proceeding at a wide variety of stages of research and development. Implantable drug delivery systems, for example, were

Table 4
OTHER ADVANCED DRUG DELIVERY SYSTEMS

Advanced polymers	Drugs are embedded in semiporous polymer membranes or entrapped in polymeric pockets or microchannels
Bioadhesives	Transdermal-like patches or tablets which adhere to the mucus linings in the mouth, gastrointestinal tract, nose, or vagina
Chewing gum forms	Particles of active drugs are isolated from flavored chewable substances and released immediately upon chewing
Electrostatic	Drugs are positively charged in a system negatively charged resin, prompting osmotic delivery through a microporous matrix
Implants	Drug delivery devices which are surgically implanted under the surface of the skin
Infusion	Portable mechanical devices designed for the intravenous administration of drugs in liquid form and other solutions
Macromolecular form	Polymer systems which are fabricated in such a way that proteins, enzymes, and other macromolecules can be released for a long period of time
Modulated release	Polymer-magnetic systems which release drugs at a base ratio which can be increased through the use of oscillating magnets
Osmotic systems	Administration forms wherein a drug in high concentration is separated by a semipermeable membrane from the same drug in low concentration
Pendent chains	Drugs which are covalently bonded to the backbone of a polymer and released upon degradation or enzymatic reaction
Pro drugs	Inactive compounds which are transformed by the body into active drugs
Rate limiters	Polymeric membranes which control the rate at which drugs are released from drug reservoirs
Soft drugs	Drugs which are absorbed through the skin, then become liquid carriers
Ultrasound systems	Topical drugs delivered via ultrasonic vibrations into sebrum pores
Zip-coded systems	Drugs encapsulated in liposomes may emulate a bacteria or virus; when the immune system attacks and consumes the liposome, the drugs are released
Zipnotics	Odorless, nonirritating polymer drug complex expressed via vapors through the lungs

once simply highly compressed drug pellets whose release rate was dependent on their surface area and solubility in body fluids. Early experiments with silicone rubber capsules showed promise and with the advent of the polymeric matrix, research in implant technology began in full force. Research in implantable drug delivery systems currently includes insulin delivery systems, levonorgestrel delivery systems, naltrexone delivery systems, and a host of others.[44]

Hsieh et al. have designed a system to release drugs at a controlled rate which can be modulated in order to achieve maximum therapeutic effectiveness.[45] Testing of a hemispherical magnetic device demonstrates its capacity to release drugs at a baseline controlled rate and, further, to rapidly increase this rate in response to exposure to an oscillating magnet. This system is the first documented which renders feasible the administration of drugs at a controlled, yet alterable, sustained release rate.

D. Drug Carriers and Drug Targeting

Interest in new drug delivery systems stems from the recognition that different drug therapies require drugs to be delivered to different areas of the body (see Figure 4). The systems listed in Table 5 provide a wide assortment of choices as to the location of the body wherein a drug will be administered. However, some researchers still object to the constraints involved in such a choice. They seek a universal drug delivery system, one which will automatically deliver a given drug to the afflicted area of an organism without its being applied there directly. Such a system would hold the toxic effects of therapeutically effective drugs to an absolute minimum. It would also allow higher concentrations of drugs to be delivered, thus maximizing therapeutic effectiveness. Often referred to as "magic bullets", these drug delivery systems allow practitioners a wide variety of options in treating diseases

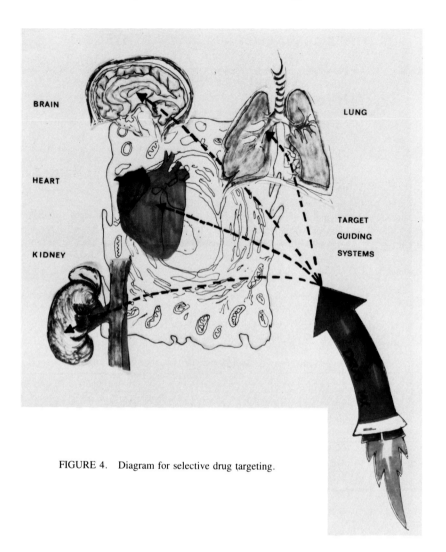

FIGURE 4.   Diagram for selective drug targeting.

which affect only a specific area of the body. Such systems are classified as "selective drug targeting" systems and will be covered in more detail in later chapters and in other volumes.[46]

IV. CONCLUSIONS

This chapter has made note of the fact that the decline in new drug development has spawned a rapidly growing drug delivery systems industry. Innovative new drug delivery systems produced in this new wave of research are of significant import to generate public excitement over pharmaceutical products. Convenient and effective delivery systems represent the future of modern drug therapy. Companies will soon recognize that public awareness of this change of focus will affect consumer view of their products.

This book provides a source of information on innovative drug delivery systems ranging from microcapsules to implants. Chapters include details on coacervation methods for preparing microspheres, multiple lamination techniques for transdermal patches, the production and fragmentation of antibodies, the coupling of antibodies with existing drugs, and steri-

Table 5
DRUG-TARGETING DELIVERY SYSTEMS

Brain-targeting	A drug quaternary salt complex is introduced in a lipid-soluble form; it is rapidly oxidized in vivo and quickly eliminated, except for that substance which passes the brain barrier in lipid-soluble form, since the ionic, hydrophilic complexes resulting from oxidation are trapped in the brain by that barrier. Enzymes must remove the drug from the carrier before either can escape through the brain barrier.
Dextran systems	Microspherical carriers block the arterial tree of the liver or kidney, resulting in increased concentrations of drug substance in those areas. This technique can be varied simply by increasing the size and number of microspheres delivered.
Iontophoresis forms	Electronic currents are used to guide drugs to target locations deep in tissues.
Lisotropic forms	Drug-containing polymers are covalently bonded to a special digestible cell carrier, usually targeted at the liver or kidney.
Magnetic systems	Drugs are processed with ferro fluids or ferric sulfate, capacitating externally controlled methods of targeting through the use of magnets.
Monoclonal antibody systems	Specific antibodies are produced by fusing myeloma cells with lymphocytes from a spleen immunized with a particular antigen. There are three distinct types of monclonal antibodies: (1) those which are a therapeutic agent in themselves; (2) those which are coupled to a drug entity, which is delivered to the location of the targeted antigen; and (3) those which are bonded to radioactive isotopes, which are released at the site of the targeted antigen

lization procedures relevant to all controlled release products. (A complete listing of the technology documented in this book is available in the Table of Contents.) This book is designed to present innovations in drug delivery from a point of view relevant to the fabrication technologies industry. It endeavors to create a bridge between advances in R & D manufacturing prospects for the host of new drug delivery systems which currently exist in various stages of development.

REFERENCES

1. **Bindra, J. S. and Lednicer, D., Eds.,** *Chronicles of Drug Discovery,* Vol. 1, John Wiley & Sons, New York, 1982, viii.
2. **Taber, B. Z., Ed.,** *Proving New Drugs, A Guide to Clinical Trials,* Geron, Los Altos, Calif., 1969, xvii.
3. **Lasagna, L. and Wardell, W. M.,** The rate of new drug discovery, in Drug Development and Marketing, Helms, R. B., Ed., American Enterprise Institute for Public Policy Research, Washington, D.C., 1975, 159.
4. **Lasagna, L. and Wardell, W. M.,** The rate of new drug discovery, in Drug Development and Marketing, Helms, R. B., Ed., American Enterprise Institute for Public Policy Research, Washington, D.C., 1975, 160.
5. **Lasagna, L. and Wardell, W. M.,** The rate of drug discovery, in Drug Development and Marketing, Helms, R. B., Ed., American Enterprise Institute for Public Policy Research, Washington, D.C., 1975, 155.
6. **Clymer, H. A.,** The economic and regulatory climate: U.S. and overseas trends, in Drug Development and Marketing, Helms, R. B., Ed., American Enterprise Institute for Public Policy Research, Washington, D.C., 1975, 137.
7. **Katz, M.,** The birth pangs of a new drug, *Drug Cosmet. Ind.,* p. 40, October 1980.
8. **Wardell, W. M. and Lasagna, L.,** Regulation and Drug Development, American Enterprise Institute for Public Policy Research, Washington, D.C., 1975, 77.
9. **Chien, Y. W.,** Logics of transdermal controlled drug administration, *Drug Dev. Ind. Pharm.,* 9, 499, 1983.

10. **Clymer, H. A.,** The economic and regulatory climate: U.S. and overseas trends, in Drug Development and Marketing, Helms, R. B., Ed., American Enterprise Institute for Public Policy Research, Washington, D.C., 1975, 141.

11. **Clymer, H. A.,** The economic and regulatoy climate: U.S. and overseas trends, in Drug Development and Marketing, Helms, R. B., Ed., American Enterprise Institute for Public Policy Research, Washington, D.C., 1975, 145.

12. **Chien, Y. W.,** Logics of transdermal controlled drug administration, *Drug Dev. Ind. Pharm.,* 9, 498, 1983.

13. **Feldman, E. G.,** New drug development costs, *J. Pharm. Sci.,* 70, 1, 1981.

14. **Feldman, E. G.,** Coming of age for device technology, *J. Pharm. Sci.,* 73, 713, 1984.

15. **Levy, R. A.,** Weighing the costs of novel drug delivery, *Pharm. Technol.,* 9, 16, 1985.

16. **Dickinson, J.,** Washington report, *Pharm. Technol.,* 9, 18, 1985.

17. **Dickinson, J.,** Washington report, *Pharm. Technol.,* 9, 19, 1985.

18. **Dickinson, J.,** Washington report, *Pharm. Technol.,* 9, 20, 1985.

19. **Feldman, E. G.,** New era for non-prescription drugs, *J. Pharm. Sci.,* 72, 1373, 1983.

20. **Ballard, B. E.,** Prolonged-action pharmaceuticals, in *Remington's Pharmaceutical Sciences,* 16th ed., Osol, A., Ed., Mack, Easton, Pa., 1980, 1596.

21. **Ballard, B. E.,** An overview of prolonged action drug dosage forms, in *Sustained and Controlled Release Drug Delivery Systems,* Robinson, J. R., Ed., Marcel Dekker, New York, 1978, 7.

22. **de Duve, C.,** Foreword, in *Drug Carriers in Modern Medicine,* Gregoriadis, G., Ed., Academic Press, New York, 1979, x.

23. **de Duve, C.,** Foreword, in *Drug Carriers in Modern Medicine,* Gregoriadis, G., Ed., Academic Press, New York, 1979, ix.

24. **de Duve, C.,** Foreword, in *Drug Carriers in Modern Medicine,* Gregoriadis, G., Ed., Academic Press, New York, 1979, x.

25. **Kondo, A.,** *Microcapsule Processing and Technology,* Marcel Dekker, New York, 1979, 27.

26. **Fanger, G. O.,** What good are microcapsules, *Chem. Technol.,* 4, 397, 1974.

27. **Madan, P. L.,** Methods of preparing microcapsules: mechanical methods, *Pharm. Technol.,* p. 26, August 1978.

28. **Madan, P. L.,** Methods of preparing microcapsules: mechanical methods, *Pharm. Technol.,* p. 28, August 1978.

29. **Goodwin, J. T. and Somerville, G. R.,** Microencapsulation in physical methods, *Chemtech,* p. 625, October 1984.

30. **Lim, F. and Moss, R. D.,** Microencapsulation of living cells and tissues, *J. Pharm. Sci.,* 70, 351, 1981.

31. **Lim, F. and Moss, R. D.,** Microencapsulation of living cells and tissues, *J. Pharm. Sci.,* 70, 354, 1981.

32. **Lim, F. and Moss, R. D.,** Microencapsulation of living cells and tissues, *J. Pharm. Sci.,* 70, 353, 1981.

33. **Chang, T. M. S.,** *Artificial Cells,* Charles C Thomas, Springfield, Ill., 1972, 180.

34. **Karim, A.,** Transdermal delivery systems, *Pharm. Technol.,* 7, 30, 1983.

35. **Karim, A.,** Transdermal delivery systems, *Pharm. Technol.,* 7, 29, 1983.

36. **Chien, Y. W.,** Logics of transdermal controlled drug administration, *Drug Dev. Ind. Pharm.,* 9, 512, 1983.

37. **Cabana, B. E.,** Regulatory considerations in transdermal controlled medication, *Drug Dev. Ind. Pharm.,* 9, 712, 1983.

38. **Skelly, J. P., Barr, W. H., Benet, L. Z., Doluisio, J. T., Goldberg, A. H., Levy, G., Lowenthal, D. T., Robinson, J. R., Shah, V. T., Temple, R. J., and Yacobi, A.,** Report of the workshop on controlled release dosage forms: issues and controversies, *Pharm. Res.,* 4(1), 75, 1987.

39. **Cabana, B. E.,** Regulatory considerations in transdermal controlled medication, *Drug Dev. Ind. Pharm.,* 9, 717, 1983.

40. **Fara, J. W.,** Short and long term drug delivery systems, *Pharm. Technol.,* 7, 37, 1983.

41. **Kligman, A. M.,** A biological brief on percutaneous absorption, *Drug Dev. Ind. Pharm.,* 9, 535, 1983.

42. **Kligman, A. M.,** A biological brief on percutaneous absorption, *Drug Dev. Ind. Pharm.,* 9, 537, 1983.

43. **Stoughton, R. B. and McClure, W. O.,** Azone: a new non-toxic enhancer of cutaneous penetration, *Drug Dev. Ind. Pharm.,* 9, 725, 1983.

44. **Sanders, H. J.,** Improved drug delivery, *Chem. Eng. News,* 63(13), 30, 1985.

45. **Hsieh, D. S. T. and Langer, R.,** Magnetic modulation of insulin release in diabetic rats, 9th Int. Symp. Controlled Release of Bioactive Material, Ft. Lauderdale, Fla., 1982.

46. **Gregoriadis, G., Ed.,** *Drug Carriers in Biology and Medicine,* Academic Press, New York, 1979.

Chapter 2

LATEX SYSTEMS FOR CONTROLLED RELEASE COATING

Carl R. Steuernagel, Aracelis Maria Ortega, and Jesse Wallace

TABLE OF CONTENTS

I. INTRODUCTION

Various strategies for design and development of prolonged-action dosage forms must achieve an improved state of disease management. This is the purpose of controlled release technology in medicine. Dosage form design based on biopharmaceutical principles and the use of fabrication techniques to maintain consistency in drug delivery profiles are exciting new fields of inquiry, made possible by recent discoveries about the fate of drugs in the body.

Controlled release is accomplished through the maintenance of precise drug blood levels within a close therapeutic range for a given drug and an appropriate route of administration. The design of sustained or controlled delivery systems for drugs to be taken orally must take into account a number of different biological components, including enzymes, gastric emptying times, pH, and partition coefficients across biological membranes and at the site of drug absorption. It follows then, that fabrication of drug delivery vehicles (and important design criteria) must be viewed in a holistic fashion, so that the biological activity of a drug is enhanced by adduct over the total duration of transit, absorption, and elimination from the body. Maintaining the desired therapeutic blood levels over (realistically) an 8 to 12-hr transit time in the GI tract presents a formidable design challenge to scientists aiming at an idealized constancy or zero-order rate kinetics of drug delivery from initial (loading) dose through specific sustaining levels (rate constant) and finally elimination.

There are difficulties in establishing methods for the quantitative prediction of diffusion coefficients for complicated organic molecules in polymers. There is wide variation in drug absorption kinetics, nonequilibrium transport parameters, and, in human metabolic activity, GI transit times, physicochemical drug properties, dosage levels, etc. This results, not surprisingly, in a great number of separate and precise (tailored) fabrication schemes for solid dosage delivery vehicles. In fact, there have been some 274 U.S. patents issued from 1974 to 1980 which deal in fabrication techniques ranging from matrix multicoating pellets and laminated tablets to lipid mixtures and expanding hydrogels. Materials range from zein, beeswax, and acacia to methylmethacrylate, 2-hydroxyethylmethylmethacrylate (HEMA), and hydrogels based on substituted methylenediamine cross-links. Different theoretical considerations in design pertain to each of the various classes of drug delivery systems: (1) matrices, (2) membrane-controlled reservoir systems, (3) bioerodible systems, and (4) pendent chain systems.

Release kinetics from inert matrix systems for polyvinyl chloride sulfinalamide wax are described by Schwartz et al.,[1] whereas Zentner et al.[2] have investigated diffusion through hydrogel membranes for progestin. Most of the disperse-type matrix devices have been prepared from polydimethyl siloxane.

Bioerodible systems have therapeutic importance for implants and their erosion involves hydrolysis and cross-link cleavage of high molecular weight water-soluble polymers. Although beyond the scope of this chapter, poly(ortho)esters have been identified for the release of naltrexone and contraceptive steroids by Benagiano and Gabelnick[3] and Heller et al.[4]

Pendant chain systems involve the binding of various agents via degradable linkages to different polymer systems. Enzyme-catalyzed cleavages in the body occur so as to release the active agent over a predetermined time frame. Hydrolysis kinetics are more complex than those of diffusional systems and are defined by Agarwal and Dhar[5] for steroidal polypeptides and Sparer et al.[6] for glycosaminoglycan prodrugs.

Good and Lee investigated membrane-controlled reservoir systems and by way of introduction to their chapter on this subject observed, ''Of all the new controlled release technology available today, the use of membranes and their ability to maintain constancy in the drug delivery profiles of compounds incorporated therein shows the most promise for providing the greatest impact on modern therapeutics. However, this technology is at the same time new and old. It is new to its application in pharmaceutics, yet old in the sense that diffusion across membranes has been known and characterized in the basic sciences for many years.''[7]

The latex system employs a continuous film membrane of amorphous and semicrystalline polymers above their glass transition temperatures: plasticized submicron dispersions which have coalesced to form a largely homogeneous film surrounding a drug core or reservoir. The kinetics of drug release are thought to be largely diffusion-dominated and transport processes are described by a single-component diffusion process based on Fick's Law or by multicomponent approximations based on the mathematics of nonequilibrium thermodynamics. In any case, transport is driven by the maintenance of a concentration gradient across the membrane from reservoir to sink.

II. THE LATEX

Latex polymer films offer a very useful new tool to transform water-insoluble polymers into water-based coating materials: finely divided submicron dispersions of such polymers as ethylcellulose, cellulose acetate phthalate, polyvinyl acetate phthalate, and methylmethacrylate copolymers have been prepared by emulsification technique for film application to solid dosage forms. These polymers possess characteristic properties of solubility and permeability in the digestive enzymes of the GI tract, depending upon the content of acidic, basic, and hydrophilic groups in the polymer.

Trends toward increased sophistication in dosage design, coupled with regulatory legislation having a bearing upon solvent emissions, have initiated an industry structural change where performance ''rate-limiting'' polymers are now being evaluated from aqueous vehicles for sustained and controlled release solid dosage forms. By pharmaceutical standards, the market for these performance film-coating excipients has become more demanding in the past 5 years, impacted by (1) patient demand for increased specificity in drug targeting, extended dosage intervals, and minimization of unwanted side effects in dosage regimens and (2) increased pharmaceutical development budgets devoted to a proactive evaluation of aqueous film alternatives. Physical and mechanical properties, as well as the chemical properties of latex polymers and the drug itself, are now combined to meet the specific rate and duration of drug delivery required by the patient. All commercially available latex systems operate on a film membrane-reservoir approach in which release kinetics are controlled over time as the drug is transported through the film by permeation, solution diffusion, and convection through micropores. Fabrication techniques involve physical application (air atomization) and aqueous film coating by fluidized particulates containing the drug, both granules and Spansule® nonpareil seeds.

A. Background

The use of natural and synthetic polymers as coatings in the pharmaceutical industry was originally for enteric coating purposes, utilizing cellulose acetate phthalate (CAP) and shellac. The first industrial applications of nonenteric film coating occurred late in 1953 with the patented Abbott Filmtab® tablet coatings. A rapidly growing interest in film coating as a new coating process followed, not only with rapidly soluble, protective, and enteric coatings, but also with controlled release pharmaceutical coatings.[8,9] A major research effort directed toward the improvement and understanding of the process and the identification of new film coating materials followed during the 1960s with the production of several hundred patents and many papers on the subject, as nearly every major drug company worked to develop its own system. Film-coating polymers were usually applied to tablets or granules from organosols, i.e., from solutions of plasticized polymers in organic solvents.

Three main drawbacks exist for the film-coating process as it has evolved over the last two decades: (1) the toxicity of the vapors and/or flammability of the organic solvents involved in the process, (2) environmental contamination resulting from the solvent vapors that are usually discharged to the atmosphere on the evaporation of the film-coating solvents, and (3) the question of the availability and the high and rising cost of the organic solvents. A further major problem has been the high cost of the solvent recovery systems necessary to prevent discharge of the solvent vapors to the atmosphere and the pressure of state and federal legislation or regulations to prohibit such discharges. The rapid increase in solvent prices in recent years, in addition to increasingly restrictive antipollution legislation which regulates the amount of organic solvent that can be emitted to the atmosphere, has led researchers to seek new water-based polymers or reexamine older ones for general-purpose film coating.

Water as a coating medium has a number of practical advantages. Its use involves no fire risk or explosion hazard, it is the most inexpensive solvent, and it is free of toxic effects and atmospheric pollution. The use of aqueous polymer solutions for film-coating purposes, however, is limited to low-solid content, since the viscosity of such solutions rises sharply with an increase in concentration and/or molecular weight of the polymer. As a consequence of this relatively low polymer concentration, a number of separated layers of polymer must be built up in order to obtain a coating of adequate thickness for surface protection. This, in addition to the slow rate of evaporation of water and the relatively large amount of water to be removed, may result in a very long processing time. Furthermore, water-sensitive drugs are prone to hydrolysis due to the long exposure of the drug to the effects of water, and many tablets must be sealed with water barriers, such as a shellac coating, prior to aqueous film coating.

Another major restriction of aqueous film coating is that the technique is currently limited to a very few water-soluble materials, which precludes enteric or controlled release coating. In addition, the strongest films result from the use of longer chain, higher molecular weight polymers, which cannot reasonably be employed by current aqueous coating methods. An alternative procedure that would revolutionize aqueous film coating would be the application of the coating material from a dispersed system, such as a finely divided, colloidal polymeric dispersion in water or a latex system. This technique would permit the incorporation of water-insoluble polymers as well as allow the use of much higher concentrations of polymers for a more efficient and low-cost film-coating process. The limitation to this approach was that prior to 1980 there were no commercially available latices which contain polymers approved for direct addition to foods or drugs in the U.S.

The technique of emulsion polymerization, originally developed in the rubber industry,[10] provides the paint and adhesives industries with aqueous dispersions of polymers as synthetic latices which serve as the vehicles for a wide range of latex paints, caulking compounds, sealers, adhesives, and similar products. The coating industry has accepted the challenge of replacing organic solvent-based systems with solvent-less coatings. Recent estimates[11] indicate that water-based coatings, which represented only 4% of the global industrial coating market in 1973, increased to 40% by 1979 and will reach as high as 60% by 1988. This potential market for water-based coatings justifies the efforts which are being made to develop and improve this class of coatings[12] and overcome all possible problems derived from the shift from organic solvent-based systems to water-based coatings.

The European Pharmaceutical Industry has introduced to the market[13] the use of two aqueous synthetic polymer dispersions for the coatings of pharmaceutical dosage forms. These dispersions, which are available under the trade names of Eudragit® L-30D and Eudragit® E-30D, have been used for a number of years. Eudragit® L-30D consists of a polyacrylic methacrylate copolymer ester which is resistent to gastric fluid, but which is soluble in a weakly basic medium due to some residual acrylic acid/methacrylic acid functionality. Eudragit® E-30D has the same basic structure of polyacrylate-methylacrylate ester, but with little or no carboxylic acid functionality, which can be formulated for rapidly disintegrating films by combining it with additives that either dissolve or swell in the stomach, or which can be used without additives for controlled release coatings.

Conventional emulsion polymerization is concerned with the production of latices from aqueous dispersions of their corresponding monomers[14] and thus it is limited to relatively water-immiscible monomers polymerizable by vinyl-addition polymerization. Latices of other polymers which cannot be prepared by emulsion polymerization from their monomers, such as polyethylene, polyurethanes, ethyl cellulose, and others, can be produced by special emulsification techniques in water, organic solvent solutions, or melts of the polymers previously prepared by other polymerization processes using conventional emulsifiers and emulsification techniques.[15] The preparation of latices by emulsification techniques applied to previously obtained polymers has provided a useful new vehicle to transform water-insoluble polymers into water-based coating materials.

It is of particular importance that polymers intended to be used for the film coating of sustained release pharmaceutical dosage forms must be physiologically safe and have a clear status as permissible food additives. Polymers such as ethyl cellulose, CAP, and poly(vinyl acetate) phthalate, which are FDA approved materials for direct addition to food[16] and for pharmaceutical film coatings, may be transformed into latices by the new emulsification techniques.

This chapter focuses on the preparation and controlled release film application of colloidal or near-colloidal polymer dispersions, including two insoluble cellulosic polymers which have been used for many years in the pharmaceutical industry as film coatings for tablets and nonpareils. The polymers included ethyl cellulose, which is usually combined with appropriate plasticizers in organosol preparations for either fast or controlled release film coatings; the solubility of the film depends on the proportions of the polymer combinations. The other polymer is CAP, which has been used for over 27 years[17] and is currently the most popular enteric coating.

B. Latices: Types and Definitions

The word latex is defined as the viscid, milky juice secreted by the laticiferous vessels of several plants and trees.[18] Currently the word latex is also used to refer to aqueous colloidal dispersions of synthetic polymers as prepared by emulsion polymerization.[19] According to their origin, latices have been classified into three categories:[20]

1. Natural latices occur as the natural products of certain plants and trees. The most important natural latex is obtained from the *Hevea brasilensis* tree,[21] which is a colloidal suspension of rubber particles stabilized by protein.
2. Synthetic latices are the products obtained by the process of emulsion polymerization. The latices are prepared from their monomers and consist of submicroscopic spherical polymer particles colloidally suspended in water.
3. Artificial latices are colloidal dispersion of polymers prepared by direct emulsification of the bulk polymer in an aqueous medium.

Regardless of their origin, latices can be generally defined as colloidal dispersions of submicroscopic spherical particles of polymers, ranging in size from 0.01 to 0.1 μm. These dispersions are milky, opaque, and fluid, even at high solid content.[22] Latices are characterized by their low viscosity, which is independent of the molecular weight of the polymer.

C. Preparation of Synthetic Latices by Emulsion Polymerization

For industrial purposes latices are generally produced by emulsion polymerization. A monomer or mixture of monomers is emulsified in water and polymerization is induced in the aqueous phase by an initiator. In emulsion polymerization surfactants play a very important role.[23] Their absorption at the interface lowers the interfacial tension between the dispersed and continuous phases and surrounds the particles with a firmly bound water envelope, stabilizing the emulsion against coagulation. The absorbed layers of amphipathic surfactants are strongly oriented in such a way that their hydrophilic polar heads are pointing into the continuous phase, while the hydrophobic nonpolar tails are anchored in the dispersed phase.

A second important function of surfactants in emulsion polymerization is the solubilization of the water-insoluble monomer. The surfactant forms aggregate structures or micelles, with the hydrophobic ends attracted toward the center of spherical or laminar micelles and with the polar moiety oriented toward the continuous phase. The nonpolar portion of the micelles is able to solubilize a considerable amount of materials which are water insoluble but oil soluble. The water-insoluble monomer then enters the micelles and becomes solubilized inside, until the solubility limit is approached and emulsification occurs.

D. Mechanism of Emulsion Polymerization

Emulsion polymerization is free-radical addition polymerization. Polymerization nuclei begin to form in the aqueous phase and free radicals, which are generated at the expense of the initiator, absorb monomer molecules and grow into latex particles. Several mechanisms which are involved, possibly simultaneously, have been proposed in emulsion polymerization, distinguished by the locus of polymerization initiation.[24]

Initiation in emulsifier micelles — Radicals are generated in the aqueous phase, enter the monomer-swollen emulsifier micelles, and rapidly polymerize the solubilized monomer inside the micelles forming a monomer-polymer particle.[25,26] Particle initiation continues until all the micelles have captured a radical or are formed to disband. The locus of polymerization initiation is the monomer-swollen micelles.

Initiation in the aqueous phase — Radicals generated in the aqueous phase add solute monomer molecules to form oligomeric radicals, which precipitate at a critical chain length to form a stable primary latex particle.[27] Latex particles are formed by homogeneous nucleation in the aqueous phase. Polymerization proceeds at the particle/water interface,[28] resulting in the formation of macromolecular conformations whose nature depends on their structure and on the relation between the molecular interaction energy and the hydration energy of the polar regions. The more hydrophilic the monomer, the lower the monomer-water interfacial tension and the smaller the concentration of absorbed emulsifier. The functional groups of the hydrophilic monomers stabilize the latex particles. The kinetics of the process is strongly affected by the interactions of the aqueous phase with the monomer.

Initiation in the monomer droplets — The two theories mentioned above agree in that the monomer droplet is merely a storehouse from which monomer moleucles diffuse into the aqueous phase. According to Harkins,[25,26] the average droplet size cannot compete with the much smaller micelles in capturing radicals. Initiation of polymerization in monomer droplets is limited by the size of the droplet; therefore, if they are small enough, they can compete with the aqueous phase of micelles as a locus for initiation. Vanderhoff et al.[29] demonstrated the initiation in droplets in inverse emulsion polymerization. Ugelstad et al.[30] postulated that a reduction in the particle size makes the monomer competitive in capturing radicals generated in the aqueous phase, increasing their importance as significant loci for the polymerization initiation. The increase in surface area of the droplet due to the reduction in the average particle size increases the amount of emulsifier adsorbed on the droplet surface and reduces the emulsifier concentration available for particle initiation in the aqueous phase.

Initiation in adsorbed emulsifer layers — A Russian scientist, Medvedev,[31] postulated that the most important requirement for polymerization initiation is the stabilization of the total surface by adsorbed emulsifier. The adsorbed emulsifier layer, whether micelles, monomer droplets, or monomer-swollen particles, is the principal locus for particle initiation. The mathematical treatments of these theories explaining the kinetics and the factors governing the polymerization rates will not be discussed here, since they are beyond the scope of this survey. Information about this theoretical aspect of emulsion polymerization can be found in the papers of Smith and Ewart,[32] Harkins,[25,26] Medvedev,[31] and Yeliseyeva.[28]

E. Preparation of Latices by Emulsification

Other classes of resins which cannot be produced by emulsion polymerization may be prepared in emulsion form by postemulsification of a completely reacted polymer. There are cases in which an emulsion may be prepared by the emulsification of either a resin solution in a solvent or the resin directly, if the viscosity is sufficiently low. After reviewing available literature, El-Aasser[12] concluded that the methods of preparation of artificial latices by emulsification techniques can be placed into three different basic approaches:

1. Solution emulsification:[23-36] The polymer is dissolved in a mixture of volatile solvents which are immiscible with water. This solution is then emulsified in water in the presence of suitable emulsifiers and the solvents are removed by vacuum steam distillation.
2. Phase inversion:[37-39] The polymer is first compounded with a long-chain fatty acid. The emulsion is formed by an inversion method. When the acid is thoroughly dispersed, a dilute aqueous solution of an alkali is worked slowly into the mixture at a temperature of approximately 100°F. The initial product is a water in polymer dispersion, but as more aqueous solution is added, an inversion of phases takes place and a dispersion of polymer droplets in water is obtained.
3. Self-emulsification:[40] The polymer molecule is chemically modified so that it becomes self-emulsifiable on dispersion in water or acids without the use of emulsification.

F. Mixed Emulsifier Systems

The stability of an emulsion is a function of the size of the dispersed droplets.[41] Emulsions containing larger droplets will separate rapidly, while emulsions containing small droplets will separate slowly. Schulman and Cockbain[41] reported that the use of a mixture of an anionic emulsifier and a fatty alcohol produced an emulsion of enhanced stability. They suggested that in order to produce stable emulsions, complexes of emulsifiers must be formed at the oil-water interface. The complex must consist of at least two components, one of which is appreciably soluble in water and the other appreciably soluble in oil, so that they form a closely packed film at the interface (Figure 1).

Mixtures of either anionic or cationic surfactants with fatty alcohols[42] are used in the emulsification of the styrene monomer with droplet sizes as small as 0.2 μm. Higher concentrations (15 to 25%) of mixtures of ionic surfactants with fatty alcohols are used in the production of "microemulsions"[43] which are transparent, with droplet sizes between 80 and 800 Å.

Vold and Mital[44] used a mixture of sodium dodecyl sulfate and lauryl alcohol to study the mechanism of the effect of mixed emulsifiers on enhancing the stability of emulsions. The authors concluded that the mechanism is mainly one of adsorption of the fatty alcohol at the oil-water interface to cover any bare areas not already protected by the adsorbed emulsifier (sodium dodecyl sulfate). Thus, the amount of fatty alcohol necessary to enhance the stability is that amount sufficient to complete the coverage of the oil-water interface by an adsorbed close-packed film, after which the stability of the emulsion is constant and independent of further additions of fatty alcohol.

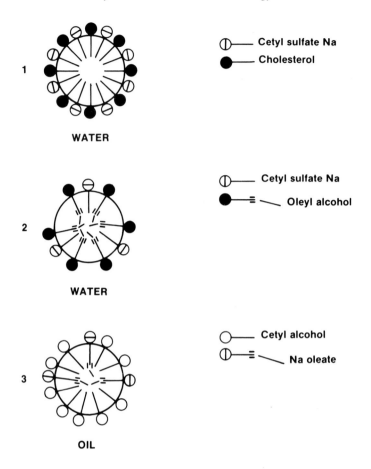

FIGURE 1.   Absorption of mixed emulsifiers at the emulsion droplet interface: (1) closely packed condensed complex, excellent emulsion; (2) no closely packed condensed complex, poor emulsion; (3) fairly poor emulsion. (From Schulman, H. J. and Cockbain, E. G., *Trans. Faraday Soc.*, 36, 651, 1940. With permission.)

G. Particle Size of Latices

Particle size and particle size distribution are very important characteristics which strongly affect the shelf stability, rheological properties, and many other physicochemical properties of any dispersed system.

The average particle diameter in natural rubber latex[45] varies with the source of the latex. The particles of a natural latex tend to be very irregular in shape and varied in size, usually ranging from 400 to 8000 Å. In contrast, synthetic latex particles are quite spherical and the latices exhibit a very narrow size distribution. Most commercial latices prepared by emulsion polymerization have average particle diameters in the range of 0.1 to 0.3 μm.

The sedimentation of particles in suspension is governed by the viscosity of the medium, the density difference between the particle and the medium, and the size of the particle itself, as stated by Stoke's Law.

It has been shown experimentally[46] that monodisperse latex particles of 1 μm in diameter will settle out upon standing within 1 to 3 months, while monodisperse polystyrene latex particles of 0.2 μm do not settle. The critical particle size below which a polystyrene latex particle with a density of 1.05 g/cm³ will not settle in a medium of 1 cP viscosity has been calculated to be 0.65 μm. Latices prepared by ordinary emulsification techniques usually have average particle diameters of 2 to 3 μm and a broad size distribution, with particles as small as 0.5 μm. This difference causes shorter shelf stability and results in poorer film properties for artificial latices as compared to synthetic latices. The use of mixed emulsifiers in the preparation of artificial latices by emulsification has been found to render improved latices with particle diameters ranging from 0.03 to 0.3 μm.[12]

H. Rheological Properties of Latex Systems

The viscosity of synthetic latices is very dependent on concentration. Dilute latices obey Einstein's equation[47] which relates the viscosity of rigid spherical particles to their concentration in very dilute suspensions:

$$\eta_r = 1 + 2.5\ \phi = \eta/\eta_o$$

where η_r = the relative viscosity of the suspension, η_o = the viscosity of the suspended particles, η = the viscosity of the medium, ϕ = the volume fraction of the sphere, and 2.5 = the shape factor. With increasing concentration, hydrodynamic interaction between the particles increases, resulting in a rapid increase in viscosity with concentration.

Many extensions of Einstein's equation[53,54] have been developed to compensate for the interaction of particles in more concentrated systems. Whereas no one equation has been found to adequately represent all systems, many of the equations can be expressed as a power series:[48]

$$\eta_r = 1 + k_1\phi + k_2\phi^2 + k_3\phi^3 + \ldots$$

where ϕ is the volume fraction, k_1 is the Einstein coefficient and k_2, k_3, etc. are higher order coefficients which account for particle interactions.

Mooney[49] derived a very interesting equation which includes a consideration of the space-crowding effect of suspended particles on each other without restrictions on concentration or particle size distribution:

$$\ln \eta_r = 2.5\ \phi/(1 - k)$$

where η_r is as defined above, ϕ is the volume fraction, and k is the self-crowding factor.

Investigations by Saunders[48] and Brodnyan[50] have shown that the viscosity-concentration dependence of latices follows Mooney's equation and appears to be applicable over a wide range of concentrations, up to a volume fraction of 0.25. Latices exhibit Newtonian flow up to an effective volume fraction of 0.25; above this concentration non-Newtonian flow of pseudoplastic materials is often observed. The viscosity of latices[50,51] was found to exhibit a particle size dependence, generally increasing with decreasing particle size. This effect of particle size is more pronounced at a high-volume fraction of latex solids.

Woods and Krieger[52] studied the rheological behavior of monodisperse latices in the high concentration range, where non-Newtonian viscosities were evaluated as a function of electrolyte level, surfactant concentration, and dimensionless shear stress, at volume fractions up to 0.50. The authors found conditions under which latices can be considered as model colloidal dispersions of rigid spheres. The specifications are

1. The particle diameters must be less than 0.5 μm.
2. Sufficient neutral electrolytes must be present to provide a compact double layer which effectively screens out interparticle coulombic forces.
3. The thickness of the adsorbed surfactant monolayer must be included as part of the radius of the sphere.

Under these conditions monodisperse latices obey a principle of corresponding rheological states, i.e.,

$$\eta_r = f(\phi \text{ and } t_r)$$

where η_r = the relative viscosity, ϕ = the volume fraction, and t_r = ta/kT (the dimensionless shear stress), t is the actual shear stress, a is the particle diameter, k is Boltzmann's constant, and T is the absolute temperature.

I. Theory of Film Formation of Latices

Some latices upon drying are transformed from milky colloidal dispersions into transparent, tough, continuous films. Others form friable, discontinuous films, i.e., a powdery material remains after solvent evaporation. The first attempt to explain the mechanism of film formation of latices was made by Dillon et al.,[53] who attributed the sintering of synthetic latices to surface tension forces and discussed the applicability of Frenkel's equation for coalescence of spherical particles by a viscous flow mechanism. Frenkel's equation,

$$\theta^2 = 3_\gamma t/2r\pi\eta_o$$

states that the degree of coalescence of rigid spheres as measured by the half angle of coalescence, θ, increases with increasing surface tension, γ; time, t; decreasing particle radius, r; and viscosity of the particles, η_o. The authors measured the angle θ as a function of radius with an electron microscope; θ^2 was found to be proportional to $1/r$, so they concluded that the coalescence of the tension of the sphere provided the necessary driving force (Figure 2). The surface pressure was calculated using the Young-Laplace equation,

$$F = \gamma(1/r_1 + 1/r_2) = 2\gamma/r$$

where r_1 and r_2 = the principal radii of curvature with values great enough to deform polymer spheres, at least at smaller particle sizes.

Brown[54] stated that the surface tension forces postulated by Dillon et al.[53] are operative but are not the principal source of energy for the coalescence of the latex particles. He proposed instead that capillary forces (F_c) resulting from the surface tension of water when evaporation has caused the formation of very small radii of curvature between the particles (Figure 3) are the main contributors of energy for the film formation process. Coalescence occurs when F_c exceeds the force resisting deformation (F_g). Voyutskii[55] postulated later that neither the surface tension nor the capillary forces alone can account for the physical properties displayed by latex films upon aging. He proposed that "autohesion", i.e., the mutual interdiffusion of free polymer chain ends across the particle-particle interfaces, along with the previously described physical forces, constitutes the most significant factor for the formation of films from latices and for the properties shown by these films.

Vanderhoff et al.[56] discussed and extended these theories. They concluded that both the surface tension and the capillary forces are active and mutually complementary during film formation of a latex, but that there are some other important radii of curvature, formed in the initial stages of latex particles coalescence, which must be taken into account. Two latex particles in a drop of water are given as a model in a detailed study of the various stages in the coalescence (Figure 4). As the water evaporates, the particles are brought together, their stabilizing layers come into contact, and further approach is hindered. The pressure forcing the particles together is increased by the further evaporation of water, i.e., by the forces arising from the water-air interfacial tension, until the stabilizing layers are ruptured and polymer-polymer contact is formed. Once this occurs, the pressure exerted upon the

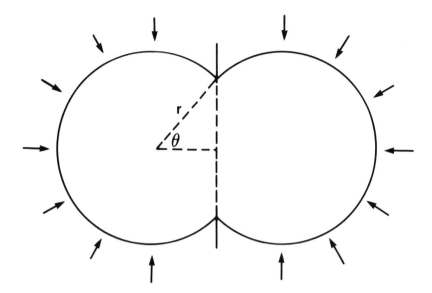

FIGURE 2. Coalescence of spheres by viscous flow caused by surface tension forces. (From Dillon, R. E., Mathenson, L. A., and Bradford, E. B., *J. Colloid Sci.*, 6, 108, 1951. With permission.)

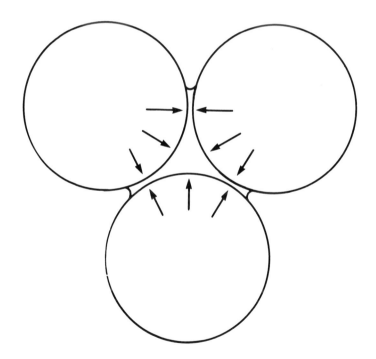

FIGURE 3. Coalescence of spheres caused by capillary forces. (From Brown, G. L., *J. Polym. Sci.*, 22, 423, 1956. With permission.)

particles is increased by the forces arising from the polymer-water interfacial tension. This initial polymer-polymer contact generates very small radii of curvature which have a strong influence on the magnitude of the surface forces exerted to cause coalescence (Figures 5 and 6).

Numerical values for the pressure exerted upon the particles were calculated using the

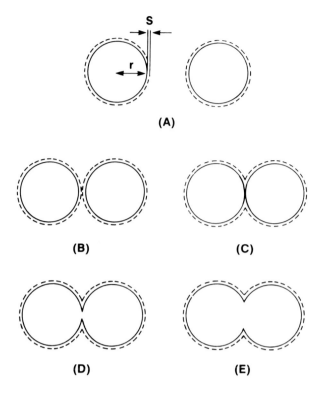

FIGURE 4.   Various stages in the coalescence of two spheres. (From
Vanderhoff, J. W., Tarkowski, H. L., Jenkis, M. C., and Bradford,
E. B., *J. Macroml. Chem.*, 1, 361, 1966. With permission.)

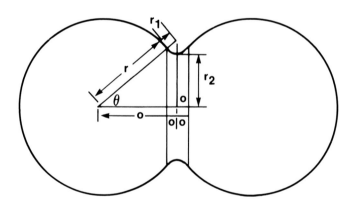

FIGURE 5.   Diagrammatic representation of the coalescence of spheres. (From
Vanderhoff, J. W., Tarkowski, H. L., Jenkis, M. C., and Bradford, E. B.,
J. Macromol. Chem., 1, 361, 1966. With permission.)

Young-Laplace equation as a function of latex particle diameter, degree of coalescence, and
interfacial tensions of both water-air and polymer-polymer interfaces. For this system the
Young-Laplace equation is

$$P = \gamma(1/r_1) - (1/r_2) + [2/\gamma]$$

where r_1 and r_2 are the important radii of curvature, r is the radius of the spheres, and γ is
the interfacial tension.

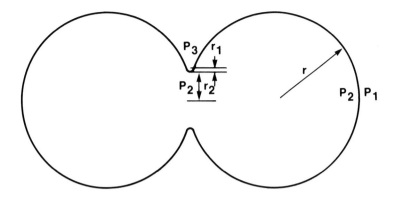

FIGURE 6. The radii of curvature involved in the coalescence of spheres. (From Vanderhoff, J. W., Tarkowski, H. L., Jenkis, M. C., and Bradford, E. B., *J. Macromol. Chem.*, 1, 361, 1966. With permission.)

Bradford and Vanderhoff[57,58] showed that a further gradual coalescence occurs in latex films which are already dry, transparent, and continuous upon aging at room temperature. Replicas of the dried films were studied by electron microscopy. The contours of individual particles could still be discerned in freshly prepared films, but coalescence gradually eliminated these contours within about 14 days. Concurrently, exudations of incompatible emulsifiers appeared on the surface of the films. As the films underwent further gradual coalescence to become more homogeneous, the incompatible emulsifiers were exuded to the surface. The authors suggested that this phenomenon may be attributable to autohesion or the action of the polymer-air interfacial tension over the very small radii of curvature of the coalesced particles.

Vanderhoff[59] reviewed the various factors that have been reported to influence the film formation of latex: polymer composition, type and concentration of plasticizer, aging, exudation of incompatible emulsifiers, and polymer structure.

Vanderhoff et al.[60] experimentally studied the drying rates of latex polymeric films. The experimental results rendered a sigmoidal curve relating the volume fraction of the polymer with time as the water evaporated (Figure 7). They divided the sigmoidal curve into three well-defined stages that were correlated with the mechanisms of film formation as follows:

1. Initially there is a constant-rate stage in which the particles are free to move with their characteristic Brownian movement and the water evaporates at the same rate as for pure water or for a dilute emulsifier solution.
2. An intermediate stage follows in which the rate of water evaporation drops off rapidly as the particles come into irreversible contact with one another, causing them to coalesce with the remaining water filling the interstices.
3. In the final stage, the rate of evaporation is much smaller and the water remaining in the film escapes by diffusion, either through capillary channels between the deformed spheres or through the polymer itself; the rate decreases very slowly with time.

J. Cellulosic Polymers

Cellulosic polymers (Structure 1) are one of the most interesting families of pharmaceutical film-forming materials. They are chemically modified natural polymers and are extremely versatile and widely used in many fields of application, including foods, plastic, textiles, pharmaceuticals, and others. These materials are all derivatives of the same natural polymer, cellulose, which is the most readily available organic raw material, since it is the chief constituent of nearly all forms of plant life. It constitutes over 98% of cotton fiber, as much as 50% of most wood fiber, and about 35% of cereal straws. Cellulose is a linear polymer

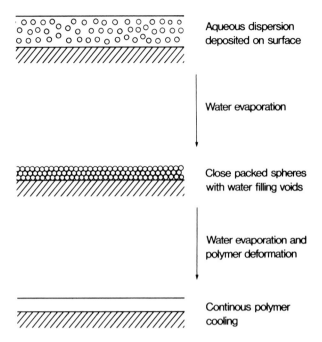

FIGURE 7.   Mechanism of latex film formation; close packing of ordered arrays of discrete spheres. (From Zdanowski, R. E. and Brown, G. L., in Proc. 44th Midyear Meet. Chemical Specialities Manufacturers Association, 1958. With permission.)

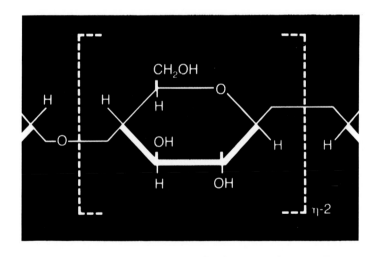

STRUCTURE 1

made up of β-glucose anhydride units, each with these hydroxyl groups, linked by primary valences through oxygen bridges.

Each natural cellulose molecule is composed of a chain of at least 3000 units.[63] Cellulose fibers are formed of groups of roughly parallel chains held together by bondings, varying from weak van der Waals forces to various degrees of hydrogen bonding. The fiber shows both crystalline areas and amorphous regions. Studies of chemical derivatives confirm the existence of one primary and two secondary hydroxyl groups for each anhydroglucose unit.

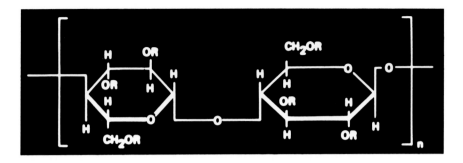

STRUCTURE 2

These hydroxyl groups provide sites for the important chemical reactions of cellulose leading to the preparation of useful derivatives.[62]

The main types of cellulose derivatives commercially available are the ethers and the esters.[61] In etherification, the hydrogen atom of the hydroxyl group is replaced by alkyl groups, the oxygen atom of the hydroxyl groups acting as a bridge. In esterification, the hydroxyl groups are replaced by ester groups such as acetate. Ethers of cellulose are organosoluble and thermoplastic, water-soluble, or alkali-soluble, depending upon the degree of structural change.[64] The etherification of cellulose usually consists of the preparation of an alkali cellulose and later reaction with etherifying reagents.[65]

The properties that best describe a particular cellulose ether are viscosity and degree of substitution (DS). DS of cellulose derivatives may be defined as the average number of molecules of substituents introduced for each anhydroglucose unit.[65]

1. Ethyl Cellulose

Ethyl cellulose (Structure 2) is prepared by the etherification of alkali cellulose with ethyl chloride, followed by the isolation, washing, and drying out of the product. The reaction is[65]

$$R_{cell}(OH)_3 \cdot 3NaOH + 2CH_3CH_2 - Cl \rightarrow R_{cell}[OH(OCH_2CH_3)] + 2NaCL$$

$$+ NaOH + 2H_2O$$

The number of moles that react varies with the degree of substitution:

1. Commercial ethyl cellulose ranges in DS from 2.2 to 2.58 (ethoxyl content from 44.0 to 49.5%). It is soluble in common organic solvents and is thermoplastic.
2. Nearly completely substituted ethyl cellulose ranges in DS from 2.6 to 2.8 (ethoxyl content from 50.0 to 52.5%). It is soluble in hydrocarbons but insoluble in oxygenated solvents. It is incompatible with commercial types of lower degrees of substitution.
3. Low-substituted ethyl cellulose ranges in DS from 0.8 to 1.7 (ethoxyl content 19 to 35%). It is water soluble.

Two substitution grades of commercial ethyl cellulose satisfy most needs.[66] One grade has a DS of 2.24 to 2.38 with an ethoxyl content of 45 to 47% and produces tough plastics with good low-temperature properties. This grade is mainly used for plastic injection molding. The other grade has a DS of 2.44 to 2.58 with an ethoxyl content of 48.0 to 49.5% and has a lower softening point, greater solubility, and good water resistance. This grade is mainly used for coating applications.[66]

Each substitution grade is produced in several viscosity types. A low viscosity is used where a high concentration in solution or a high flow is required. Higher viscosity types are used whenever strength, flexibility, or hardness are the important factors. Ethocel standard R, at all viscosities, complies with the National Formulary XIII and is listed in the Food and Drug Administration Additive Amendment as an approved food packaging material. It is also covered for certain direct food additive applications as recorded in Section 121.1087 and published in Volume 27 of the *Federal Register* 1962.

2. Cellulose Acetate Phthalate (CAP)

CAP is prepared by causing a partial acetate ester of cellulose to react with phthalic anhydride. Approximately half of the hydroxyl groups of the anhydroglucose units are acetylated and one fourth are esterfied with one of the two acid groups of phthalic acid. The other carboxyl group is free to react with alkali to form salts. This available free acid group imparts to CAP its unique properties of solubility in aqueous alkaline solutions and insolubility in neutral and acid media.[67]

CAP was first patented as an enteric coating in 1940. It is usually applied to tablets from organic solvent solutions of polymers, containing a plasticizer and, in some cases, pigments and opacifiers.[68] The type of plasticizer chosen plays an important role in determining the water vapor transmission rate, moisture absorption, and water permeation properties[69] of CAP films. The performance of the enteric coating is affected by the tablet substrate. The method of film application[70] influences the rate of moisture absorption of CAP-coated tablets.

III. KINETICS OF DRUG RELEASE

A. Theory

There is no clear-cut distinction between diffusion control and encapsulated dissolution control for the "membrane-controlled reservoir" system of drug transport across film. Early seed or granule products introduced in the 1950s employed a common procedure of coating the individual particles or sugar nonpareils (containing the drug) with varying thicknesses of slowly soluble coating material. These Spansule® sustained release dosage forms achieved a type of "pulsed dosing", where the time frame of control was a function of coating thickness and dissolution rate of the coating substance.

A slowly dissolving wax or polymer coat of carbohydrate sugars and cellulose, beeswax, various shellacs, or polyethylene glycol was applied to nonpareils or granules. A fraction of the seeds were left uncoated to provide for immediate release of the drug, or a loading dose, while the remaining seeds were split into groups of varying coating thickness. Examples of such coated drugs include primarily antihistimines, but also "pulsed-action" antihypertensives, anorexigenic agents, steroid anti-inflammatories, and phenothiazines.[71]

Lee and Robinson[71] note the theoretical distinction between controlled release methods using diffusion and those oral dosage forms relying on dissolution, where an eluting medium penetrates the slowly soluble membrane through the pores. "In practice it is common to see both mechanisms operative in a given dosage form, although one mechanism will usually predominate over the other."

Based upon general structural features, Good and Lee[7] classify three important designs for membrane-reservoir drug delivery systems:

1.   Amorphous or semicrystalline polymers: the drug diffuses down a chemical potential gradient into an external releasing medium.
2.   Glassy hydrogels: swollen polymer gels formed by imbibing water; convection and drug permeation take place by solution-diffusion method.
3.   Microporous membrane: molecules transport through pores and flow is governed by diffusion and convection.

B. Rate Limitation

Latex films applied to nonpareils (drug reservoir), via air atomization in a fluid bed apparatus, coalesce under conditions of heat to form a largely homogeneous film. The films derived in this manner exhibit rate control of drug dissolution characteristic of both (1) and (2) above; however, it may be shown that diffusion is the dominant mode of drug treatment.

Various pharmacokinetic models in the literature are proposed for different systems with the idea of identifying a rate-limiting factor. Experiments with latex films of ethyl cellulose polymers for controlled release solid dosage forms focus on the regulable transport of a drug permeating an intact film from a reservoir to an aqueous sink condition of changing pH. A drug must pass through the ethyl cellulose membrane from one compartment to another, over time, and its concentration is measured at hourly intervals in the sink dissolution medium. With a thin latex film applied to Spansule® beads it is necessary to understand both membrane-matrix-controlled (intact film) transport and diffusion through water-filled pores. Neither transport mechanism is pure and absolute and most film systems exhibit characteristics of both.

How a drug penetrates or is absorbed through a film (i.e., whether diffusion is dominated by a membrane-controlled flux or through pores) depends first on whether it is an intact or a porous film (leaching convection). In a mixed system, the rate of drug transport will show characteristics of each model, but the concept of a solute crossing a membrane inevitably depends on characteristics of the drug itself, the film properties and structure, and the product formulation and sink conditions. How fast a drug permeates the film depends on such physicochemical properties as its water solubility, polymer/water partition coefficient, ionization constant, and chemical structure. It depends further on the chemical composition of the film barrier, any ionic charge, and the morphological structure of the film.

Pharmacokinetic Models

Drug physicochemical properties	Film properties
Water solubility	Polymer substance
Polymer/water partition coefficient	Chemical composition of barrier
Dissociation constant (pK_a)	Electrical charge
Chemical structure	Morphological structure
Molecular weight	

C. Latex Film Drug Transport

In vitro dissolution results suggest that drug release through a latex film occurs by constant diffusion through the film independent of concentration as long as a concentration gradient in the coated bead or nonpareil is maintained. The latex film deposited on the bead surface regulates drug release as a linear function with time.

Latex dispersions have been applied at various film levels to two model drug systems, phenylpropanolamine HCl and theophylline. In vitro release rates of the drug from nonpareil seeds were shown to be inversely proportional to the film weight of latex dispersion (and plasticizer) applied. Rank order was recorded in all cases for families of dissolution curves, determined as a function of time and changing pH media using a rotating bottle method.

The important variables which must be characterized at the outset, and which greatly affect the release-rate profiles one might expect to observe through a latex film, relate to both the substrate or nonpareil seed and the drug, its dissociation constant, it solubility, whether it is acidic or basic, and the pH of the dissolution medium. The characteristics of the substrate which may effect the release rate are its nonpareil mesh size distribution, initial and equilibrium moisture content, static charges, and representation samples.

The surface area available for drug diffusion is also a critical variable where the mechanism

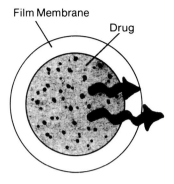

FIGURE 8.   Diffusion/dissolution
of drug through film membrane.

of drug release is diffusion controlled by a thin film membrane and the kinetics are zero order and Fickian. It is necessary in all cases to carefully characterize and monitor the mesh size and size distribution of the nonpareil seeds to be coated. Otherwise, differences in release rates might be observed for a given film thickness and plasticizer level, coated under identical fluid bed conditions. The effect of this variable is minimized by constant use of seeds of the same sieve fraction, same manufacture, and most narrow size distribution and regular geometry (spherical): phenylpropanolamine hydrochloride (PPA HCl) of an 18- to 20-mesh cut and 70 to 75% potency. In these diffusion-control models, the approach to a sustained release barrier involves enclosing the drug-impregnated nonpareil seed with a thin latex polymer coat. A constant area for diffusion together with a constant diffusional path length and constant concentration of drug are essential to achieve a constant drug release rate. Presumably, the eluting solvent permeates the coat and drug diffuses out through the coating.

Although drug release in many other systems (encapsulated diffusion control) usually involves a combination of dissolution and diffusion, with dissolution being the overriding process, observations involving the model systems presented here argue that diffusion control predominates. The reproducibility of data is strong dependent upon the starting substrate material.

The flux of drug across the membrane where a water-insoluble membrane encloses a core of drug is given by Fick's first law:

$$\frac{dM}{dt} = \frac{ADK\Delta C}{l}$$

where A = area, D = diffusion coefficient, K = partition coefficient of drug between membrane and core, l = diffusional path length (film thickness), and C = concentration difference across the film. All of the terms on the right side, ideally, should be held constant to show constant drug release over time. It is common in many oral sustained released products that one or more of the terms changes, but in this case control of the "area" term is established to approach the ideal of a reproductive, apparent zero-order release pattern (see Figure 8).

D. Drug Ionization Constant (pK$_a$)

Having isolated the critical variables relating to the nonpareil substrate, controlled closely for their effect on release rate, it is now appropriate to consider the drug itself. The drug, its pK$_a$, and its solubility will largely determine at what rate drug release will proceed from a reservoir (nonpareil seed) through the polymer film membrane. Where diffusion is the

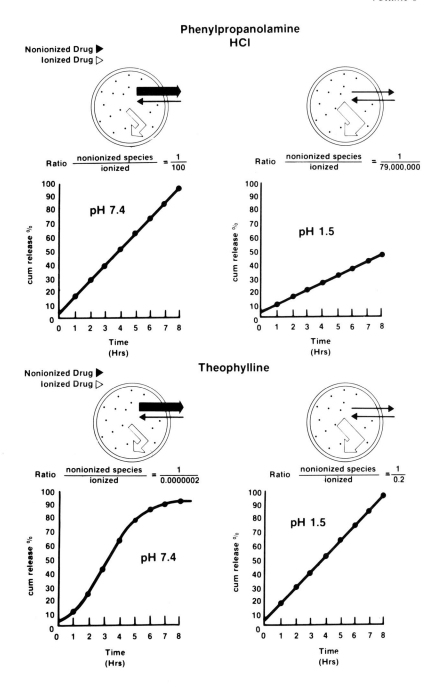

FIGURE 9. PPA HCl release through a constant latex dispersion film thickness contrasted with release patterns for theophylline.

dominant transport mechanism, it is theorized that in a given pH medium, drug release will proceed through the film at very different rates, based upon the proportion of nonionized to ionized drug molecules.

In Figure 9, PPA HCl (pK_a = 9.4) release through a constant latex dispersion film thickness is contrasted with the release patterns for theophylline (pK_a = 0.7) holding the pH medium constant for 6 to 8 hr. The illustrated effect is a rate of drug transport through a latex film for a weak base (theophylline) of low water solubility with the rate for a mildly

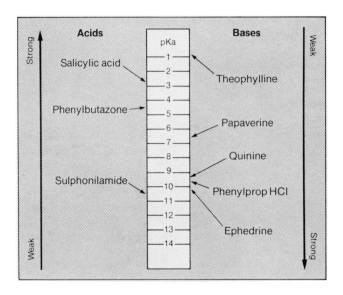

FIGURE 10.   The pK_a values of a number of familiar drugs.

strong base (PPA HCl) of high water solubility. The pK_a is the fraction of the compound present in the ionized form. Weak acids and weak bases are only partly ionized (dissociated) in water; strong acids and strong bases are completely ionized. The extent of ionization of drugs is an important determinant of their rates of passage across a film membrane by diffusion. The pK_a values of a number of familiar drugs are shown in Figure 10. Again, in a diffusion model, the extent of ionization of drugs is an important determinant of their transport rate across the latex dispersion film.

The degree of ionization depends on both the pK_a of the drug and the pH of the medium in which dissolution occurs. The theoretical basis for drug release rates of PPA HCl and theophylline through a latex film demonstrates this pH dependence and the relationship is defined by the Henderson-Hasselbalch equations.

For a base:

$$\log \frac{Ci}{Cu} = pK_a - pH$$

where Cu = nonionized form concentration and Ci = ionized form concentration.

It is possible also to contrast the proportion of nonionized to ionized molecules in an acid medium and in a basic medium. For PPA HCl ($pK_a = 9.4$) in base (pH = 7.4):

$$\log \frac{Ci}{Cu} = pK_a - pH$$

$$= 9.4 - 7.4 = 2$$

Therefore, the proportion of nonionized to ionized molecules is 1:100. In acid (pH = 1.5):

$$\log \frac{Ci}{Cu} = pK_a - pH$$

$$= 9.4 - 1.5 = 7.9$$

Therefore, the proportion of nonionized to ionized molecules is essentially 1:79,000,000.

For theophylline, with a low pK_a value, the proportions are different for acid and basic media. For theophylline ($pK_a = 0.7$) in base (pH $= 7.4$):

$$\log \frac{Ci}{Cu} = pK_a - pH$$

$$= 0.7 - 7.4 = 6.7$$

Therefore, the proportion of nonionized to ionized species is 1:0.0000002. In acid (pH $= 1.5$):

$$\log \frac{Ci}{Cu} = pK_a - pH$$

$$= 0.7 - 1.5 = 0.8$$

Therefore, the proportion of nonionized to ionized is essentially 1:0.2.

The result is a very simplified structural framework on which to base theories on the rate of drug permeation through a film formed from a latex dispersion. It is based upon the drug pK_a and this pH partition theory. Nonionized drug transports through the membrane more readily. Thus, for theophylline, a faster rate of drug transport might be expected in base (pH 7.4) and a slower rate of drug transport in acid (pH 1.5).

When PPA HCl is in acid media (pH 1.5) a much larger proportion of drug is in the ionized form and the first few hours of the dissolution test shows very little drug release. This rate then accelerates in the 6.9 and 7.2 pH media. As the proportion of nonionized drug increases, drug transport across the latex film membrane accelerates.

IV. FABRICATION TECHNIQUES

A. The Process

Experiments with an ethyl cellulose latex utilize a Wurster coating process where particulates (spherical seeds containing the drug) are fluidized on a column of air. Particles to be coated are fluidized on an upward-moving airstream. A cylindrical partition is placed around the spout formed by the airstream and a cyclic flow is induced. Since the bed of particles is fluidized, drying occurs as particles descend to the partition and, cycling in this manner, pass the spray nozzle every 6 to 10 sec, receiving additional coating.

Time and temperature are variables which are controlled carefully for the complete coalescence of a latex system. The air-atomized latex dispersion forms a fully fused integral film from ordered arrays of latex spheres deposited on the surface of cycling particulates (Figure 7).

PPA HCl presents a model system where considerable experimentation has been accomplished utilizing latex systems for controlled release. The coating of highly soluble drugs with a water-based latex poses a special challenge in that it is necessary to ''seal'' the active surface with a slow initial subcoat rate (3 to 4 mℓ/min) and the process of coalescence which all latices undergo (and which will vary release rates until complete) should be accelerated to an end point through postprocess drying at about 40°C for 30 min.

B. Substrate Nonpareil

PPA HCl nonpareils of a narrow sieve cut and distribution are employed (16 to 20 mesh) to minimize potential for drug release inconsistences as a result of sampling bias. The distribution is normal with 96% of nonpareils in the range of 840 to 1190 μm.

Seed Characterization

	Mesh	% retained on	Size (μm)
Raw seeds	16	4—6	1190
(PPA HCl)[a]	18	73—83	1000
	20	12—20	840

[a]   Potency assay (UV) = 73.7%.

The surface area available for drug diffusion is a critical variable where the mechanism of drug release is controlled by a thin film membrane coating on a bead substrate. The effect of this variable is minimized by constant use of seeds of the same sieve fraction, most narrow size distribution, and regular geometry.

C. Coating Conditions

In the Wurster process the air distribution chamber provides a baffle through which the process air is distributed uniformly to the base of the coating chamber.[72] This consists of a perforated air distribution plate, a partition, and the atomizing nozzle. There is an expansion chamber of greater diameter above in which the linear velocity of the airstream is decreased, allowing the particles to settle out. Units which will be used for development work usually have three or more plates and may have three or more partitions. Having the correct plate design and selection of the proper partition is critical to a smooth fluidization of particles.

PPA HCl seeds have been coated in a 4''/6'' Wurster column by Wheatley.[73] Process air temperatures were identified for plasticized latices of ethyl cellulose at 55 to 56°C to assure proper fluidization. At temperatures higher than 60°C it was reported that certain plasticizers (triethyl and tributyl citrates) would soften the film excessively, resulting in seed agglomeration and sticking. Dibutyl sebacate (DBS) was employed at 24% levels based on the latex polymer solids (Table 1).

Process Equipment

Column	Wurster 4''/6''
Nozzle	Spraying systems 1/4 J. Series 285070 SS
Partition	3/8'' setting
Pump	Masterflex 16 pump head

Coating Conditions

Seed load (kg)	1.0
Process air temperature (°C)	55—56
Air flow (ft³/min)	8—10
Pumping rate (mℓ/mm) (normal)	10
Pumping rate (mℓ/mm) (initial-slow)	2—3
Pump speed	Calibrate
Atomizing air (psi)	15

Coating Time (min)

10% film weight	65—73
Slow coating	29—34
Normal coating	32—41
Post-drying	30

A total of 10% film coating (theoretical) was applied to the seeds in the following fashion: a 2% seal coat applied first at slower fluid rates (2 to 3 mℓ/min) and then an additionl 8% fast overcoat (10 mℓ/min). Outlet temperatures ran between 29 and 33°C and, at the con-

Table 1
ANALYSIS (GC) OF PPA SEEDS COATED WITH LATEX FOR
PLASTICIZER PERMANENCE

DBS

			% DBS found			
				3 months		6 months
Plasticizer level (%)	% DBS theory[a]	Initial	RT	35°C	RT	35°C
24	1.94	1.73 (89.2)	1.79 (92.3)	1.85 (95.4)	1.87 (96.4)	1.83 (94.3)
30	2.31	2.02 (87.4)	2.14 (92.6)	2.14 (92.6)	2.01 (87.0)	2.09 (90.5)

Citroflex®-2 (Triethyl citrate)

			% Citroflex®-2 Found			
				3 months		6 months
Plasticizer Level (%)	% Citroflex® Theory[a]	Initial	RT	35°C	RT	35°C
30	2.31	2.15 (93.1)	—	—	1.98 (85.7)	1.77 (76.6)

Note: The numbers in parentheses refer to the percentage of theory. RT is room temperature.

[a] Calculated from % plasticizer in coating suspension and assuming a 10% coating.

clusion of the film application, the coated seeds were dried for 30 min (36 to 42°C) on a fluid column of air.

D. Plasticizers
1. Definition
Plasticizers are defined as usually nonvolatile, high-boiling substances added to polymeric films to reduce brittleness; increase toughness, strength, and tear resistance; and impart flexibility.[74] Plasticization is usually critical to good polymer films either from a polymer solution or dispersion (latex).

Softening and swelling of latex spheres aids in overcoming their resistance to deformation. More importantly, however, in latices where large surface areas exist (and therefore large surface energies), the driving force necessary to overcome the repulsive forces is known as capillarity. It is this force which fuses and deforms plasticized latex spheres into a continuous film.

A fundamental requirement of any plasticizer in a polymer system is compatiblity and permanence (the most effective plasticizers will generally have some solubility in the polymers they plasticize). To be compatible the plasticizer usually is miscible with the polymer, indicating similar intermolecular forces in the two components. Fundamental mechanical and physicochemical properties of latex films are affected by plasticization, polymer chemistry, and film additives, and this must be considered in relation to film dissolution, permeability, and diffusion properties.

Banker[74] observes that:

The plasticity of an unmodified as well as of a plasticized polymer film is related to the chemical composition of the polymer (and plasticizer) and to the arrangement, stereochemistry, and forces acting between the chain macromolecules, including intermolecular and internuclear distances, and to the effect of the regularly interposing

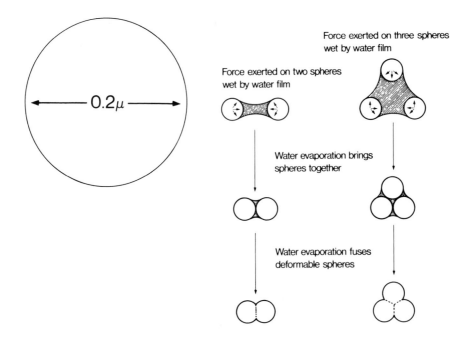

FIGURE 11.   Capillarity effects fusing discrete latex spheres as evaporation of water proceeds. Latex particles (ethyl cellulose) generally fall in the range of 0.05 to 0.3 µm with the average particle size at 0.2 µm. (From Zdanowski, R. E. and Brown, G. L., in Proc. 44th Midyear Meet. Chemical Specialities Manufacturers Association, 1958. With permission.)

plasticizer molecules within the molecular polymeric network. These relationships and effects will dictate the proportion in which the plasticizer must be used to produce the desired film properties. Cellulosic polymer films commonly require 30 to 60% plasticizer relative to polymer weight for adequate plasticization.

This is also true for latices for membrane-reservoir film formation: the total film-coating formulation of polymer and plasticizer, plus other components such as insoluble additives or surfactants to promote spreading, must be considered together as affecting the nature and properties of the film that is formed.

If the latex polymer spheres are rigid (unplasticized), such that forces of capillarity cannot produce particle deformation and therefore coalescence, a powdery discontinuous structure results. Moreover, plasticization of the polymer particles will greatly reduce the internal viscosity of the spheres and reduce the critical film-forming temperature.

2. Capillarity

Latex films involve the deposition of the film-forming polymers from a colloidal dispersion of submicron polymer spheres in water. As the water dispersion medium evaporates, the individual polymer particles, with their hundreds of polymer chains, coalesce into a continuous film. Particle deformation and fusion require a strong driving force to overcome the inherent hardness of the polymer spheres as well as their electrostatic repulsive charges. Capillarity is caused by the extremely high interfacial surface tension of water, i.e., water becomes strongly adherent to the surface it wets, and as a result of its high inherent tensile strength, surface tension, and adhesion develops a bond of great strength.

When polymer spheres are wetted (Figure 11) and the water of the enveloping wetting droplet evaporates, the spheres are pulled closer and closer together as the surrounding water film constricts. As the water continues to evaporate, at some critical point the resistance of the stabilizing layers on the adjacent spheres is overcome, and polymer-polymer contact

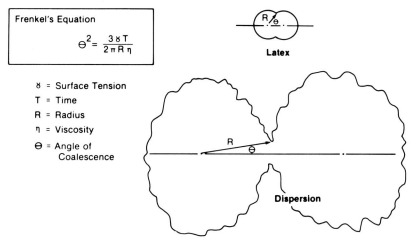

Frenkel's Equation

$$\Theta^2 = \frac{3 \, \gamma \, T}{2 \pi R \, \eta}$$

γ = Surface Tension
T = Time
R = Radius
η = Viscosity
Θ = Angle of Coalescence

Latex

Dispersion

Source: Dillion, Matheson, Bradford "Sintering of Synthetic Latex Particles" J. Colloid Sci. 6. 108. (1951)

FIGURE 12.   Comparison of viscous coalescence between latices and dispersions. (From Dillon, R. E., Mathenson, L. A., and Bradford, E. B., *J. Colloid Sci.*, 6, 108, 1951. With permission.)

occurs. Continued interfacial tension between the water and the polymer particles acts to cause a fusing of deformable spheres into a clear continuous film. Free polymer chains in the fused, deformed latex spheres (each containing thousands of polymer molecules) diffuse across what was the interfacial boundary.[53] Mutual interdiffusion of free chains then causes overall knitting of the film into one continuous polymeric sheet, and ultimate film strength, permeability, and diffusivity are developed.

Cohesion, also described by Voyutskii[55] as autohesion or self-adhesion, refers to the ability of continuous surfaces of the same material, at a molecular or supermolecular level, to form a strong bond. Banker has observed[74] that to obtain high levels of cohesion two phenomena are necessary: (1) the autohesive strength of the material, molecule to molecule, must be relatively high and (2) the contiguous surfaces of the film material must coalesce on contact.

The degree of coalescence is characterized by the angle θ (Figure 12). This coalescence improves as the surface tension (polymer/air) or interfacial tension (polymer/water) increases and the viscosity of the polymer decreases. The smaller the particles, the more complete the fusion. Frenkel's equation shows the inverse relationship between internal sphere viscosity (η) achieved through plasticization and interfacial tension (γ) or the force necessary to fuse adjacent spheres (Figure 12). Dillon et al.[53] verified that θ^2 was proportional to 1/r.

Unpublished university research and commercial applications[76] suggest that certain butylated citrates and acetylated monoglycerides at levels of 30% based on latex polymer solids might result in more consistant patterns of drug transport through the ethyl cellulose latex film. Some of the more pertinent physical data for effective plasticizers are presented in Table 2.

E. Experimental Results

Both the starting substrate (sugar nonpareils) and the time/temperature conditions of fluid bed drying are important variables to control so that consistent drug release patterns are achieved through a latex film over time. Release profiles are affected by variables relating to the drug itself, its pK_a, and solubility; the substrate, mesh size of the seeds, moisture content, and organic volatiles; and plasticizer selection and levels.

Cumulative drug release patterns have been studied by Wheatley[73] for PPA HCl nonpareils

Table 2
PLASTICIZER PHYSICAL DATA

	BP (°C)	Vapor density (air = 1)	Vapor pressure (mm Hg)	Water solubility
DBS	349	10.8	10 at 200°C	Negligible
Diethyl phthalate (DEP)	298	7.66	100 at 220°C 1 at 109°C	Insoluble (0.12 g/ℓ)
Triethyl citrate (Citroflex®-2)	294	9.7	1 at 107°C	6.5%
Tributyl citrate (Citroflex®-4)	170 (1 mm Hg)	12.4	1 at 170°C	Insoluble
Acetyl tributyl citrate (Citroflex®-A4)	173 (1 mm Hg)	14.1	0.8 at 170°C	Insoluble
Myvacet® 9-40	474	—	Nonvolatile	Insoluble

Table 3
POROSITY (MERCURY INTRUSION) OF LATEX
FILMS DEPOSITED ON NONPAREIL SEEDS;
VARIOUS PLASTICIZERS

Plasticizer	Temperature (°C)[a]	Porosity (E)	Surface area of pores (m²/g)
DBS at 24%	RT	0.019	2.19
(control)	35°	0.015	1.96
DBS at 30%	RT	0.018	2.07
	35°	0.019	1.97
Citroflex®-2 at	RT	0.019	1.94
30%	35°	0.016	1.81
Citroflex®-4 at	RT	0.020	1.90
30%	35°	0.019	1.89
Citroflex® A-4	RT	0.019	2.04
at 30%	35°	0.018	1.87
DEP at 30%	RT	0.019	1.93
	35°	0.018	1.96
Myvacet® 9-40	RT	0.020	2.05
at 30%	35°	0.019	1.81

[a] RT = room temperature.

coated to a constant film level with an ethyl cellulose latex (Table 3). All critical variables were controlled so that the plasticizer effect could be isolated and examined. As part of this study values were recorded for moisture content and plasticizer permanence in the film by gas chromatography. After fluid bed coating and 30-min column post-drying, the coated seeds with alternative plasticizers were further examined by scanning electron microscopy and mercury intrusion porosimetry to record the degree of film surface porosity and total pore volume.

Figures 13 to 22 present in vitro curves for initial, 3-month, and 6-month drug release patterns. The coated seeds were riffled (split) to obtain a statistically representative sample for dissolution testing. Drug release analysis was performed using a rotating bottle method employing changes in dissolution media pH with time according to the regimen in Table 4.

Banker and Peck[75] have investigated mixed latex and water-soluble polymer films for their dissolution properties. They found that ethyl cellulose and other water-insoluble polymers in latex form might be modified either by the addition of water-soluble, sparingly

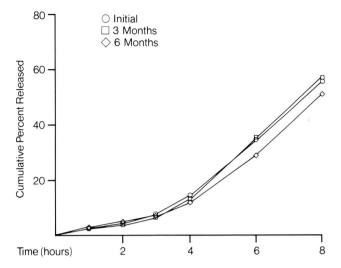

FIGURE 13.   Nonpareil seeds coated with ethyl cellulose latex; dibutyl sebacate plasticizer at 30% levels based on latex polymer solids (room temperature stability).

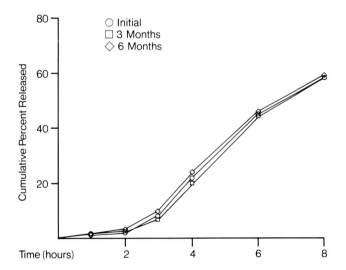

FIGURE 14.   Tributyl citrate plasticizer at 30% levels (room temperature stability).

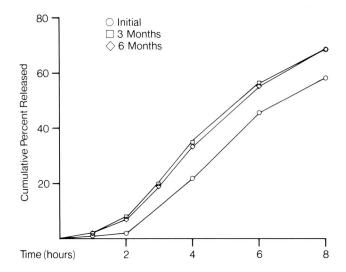

FIGURE 15.   Nonpareil seeds coated with ethyl cellulose latex; tributyl citrate plasticizer at 30% levels based on latex polymer solids (35°C stability).

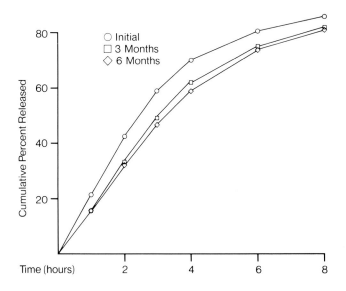

FIGURE 16.   Nonpareil seeds coated with ethyl cellulose latex; triethyl citrate plasticizer at 30% levels based on latex polymer solids (room temperature stability).

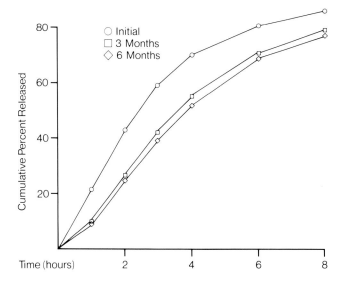

FIGURE 17.   Triethyl citrate plasticizer at 30% levels (35° stability).

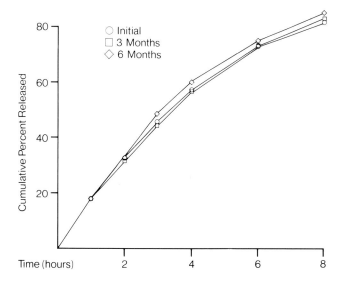

FIGURE 18.   Nonpareil seeds coated with ethyl cellulose latex; acetylated monoglyceride plasticizer (Myvacet® 9-40) at 30% levels based on latex polymer solids (room temperature stability).

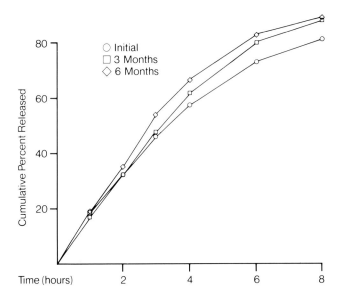

FIGURE 19.   Acetylated monoglyceride plasticizer at 30% levels (35° stability).

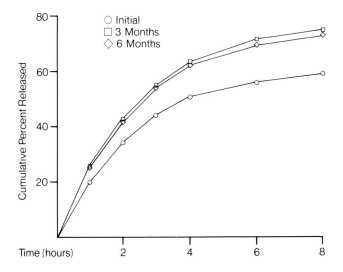

FIGURE 20.   Nonpareil seeds coated with ethyl cellulose latex; diethyl phthalate plasticized at 30% levels based on latex polymer solids (room temperature stability).

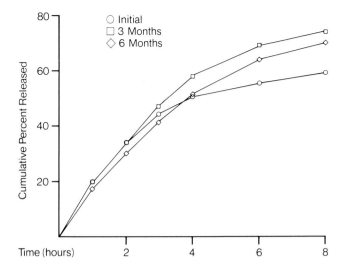

FIGURE 21.   Diethyl phthalate plasticizer at 30% levels (35° stability).

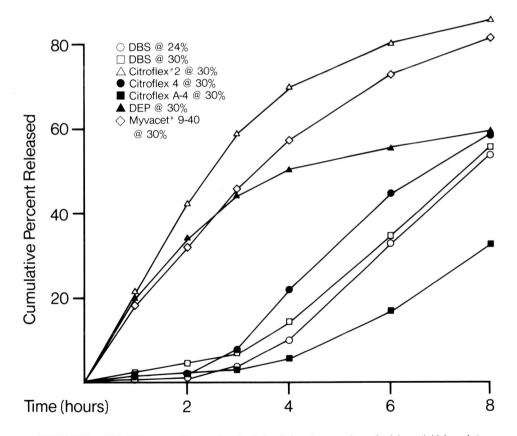

FIGURE 22.   PPA HCl nonpareils coated with ethyl cellulose latex; various plasticizers (initial results).

Table 4
CHANGING pH
DISSOLUTION (8-hr
Analysis)

Interval (hr)	Fluid	pH
0—1	Gastric	1.5
1—2	Intestinal A	4.5
2—3	Intestinal B	6.9
3—4	Intestinal B	6.9
4—6	Intestinal C	7.2
6—8	Intestinal C	7.2
Residue	Distilled water	—

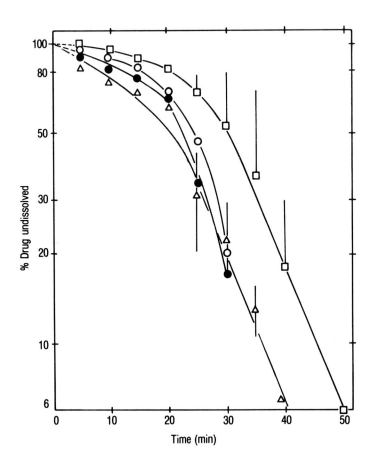

FIGURE 23.   Dissolution rates of coated and uncoated PPA tablets (the coating used was an annealed ethyl cellulose pseudolatex). △ = uncoated tablets, ● = 0.052 mm coating, ○ = 0.093 mm coating, and □ = 0.131 mm coating.

soluble, or water-insoluble plasticizers or by adding other polymers in solution to the pseudolatices (film annealing). These modifications formed films that were rapidly soluble in gastric fluid, slowly soluble in gastric and intestinal fluid, or enteric in their solubility properties.

Figure 23 presents dissolution data for an uncoated PPA tablet and for the same tablet coated with various thicknesses of an annealed ethyl cellulose latex dispersion. Each tablet contains 25 mg of drug, 175 mg of dicalcium phosphate dihydrate, 75 mg of microcrystalline cellulose, and 25 mg of corn starch with a talc lubricant. The coating used for the tablets was 400 g of a 30% solid ethyl cellulose latex, 40 g of 15-cP grade HPMC, 55 g of glyceryl triacetate, 30 g of glycerin, and sufficient water to produce 815 g of total volume. The polymer ratio of ethyl cellulose to HPMC was thus 3:1. The tablets were coated in a laboratory air-suspension tower. Over the range of coating thicknesses employed, differences in the dissolution time of the average amount of undissolved drug at any point in time were 10 min or less. The difference between dissolution half-time for the two lower coat weights and the uncoated tablets was 5 min or less. Figure 23 indicates the ability to use the ethyl cellulose latex dispersion as a water-soluble coating without substantially altering the dissolution profile of the coated tablets in comparison with the uncoated control tablets. The dissolution test employed was USP apparatus 1, employing 900 mℓ of simulated gastric fluid (without enzyme) at 37°C (± 1°C) and a speed of rotation of 100 rpm.

Other coating formulations examined for their ability to form soluble coatings on the PPA tablets included slightly lower or higher ratios (3:2 or 4:1) or ethyl cellulose latex and HPMC, as well as slightly different ratios of plasticizer and the use of propylene glycol in place of glycerin.

REFERENCES

1. **Schwartz, J. B., Simonelli, A. P., and Higuchi, W. I.,** Drug release from wax matrixes. I. Analysis of data with first order kinetics and with the diffusion controlled model, *J. Pharm. Sci.,* 57, 274, 1968.
2. **Zentner, G. M., Cardinal, J. R., and Kim, S. W.,** Progestin permeation through polymer membranes. II. Diffusion studies on hydrogel membranes, *J. Pharm. Sci.,* 67, 1352, 1978.
3. **Benagiano, G. and Gabelnick, H. L.,** Biodegradable systems for the sustained release of fertility-regulating agents, *J. Steroid Biochem.,* 11, 449, 1979.
4. **Heller, J., Penhale, D. W. H., and Helwing, R. F.,** Preparation of poly(ortho) esters by the reaction of dikentene acetyls and polyols, *J. Polym. Sci. Polym. Lett. Ed.,* 18, 619, 1980.
5. **Agarwal, K. L. and Dhar, M. M.,** Steroidal polypeptides. III. The synthesis of a α-poly-ε-deoxycholic acidamido-L-lysine, *Steroids,* 6, 105, 1965.
6. **Sparer, R. V., Ekwuribe, N., and Walton, A. G.,** Controlled release glycosaminoglycan prodrugs, in *Controlled Release Delivery Systems,* Roseman, T. J. and Mansdorf, S. Z., Eds., Marcel Dekker, New York, 1984.
7. **Good, W. R. and Lee, P. I.,** Membrane controlled reservoir drug delivery systems, in *Medical Applications of Controlled Release,* Vol. 1, Langer, R. S. and Wise, D. L., Eds., CRC Press, Boca Raton, Fla., 1984.
8. **Munden, B. J., DeKay, H. G., and Banker, G. S.,** Evaluation of polymeric materials. I. Screening of selected polymers as film coating agents, *J. Pharm. Sci.,* 53, 395, 1964.
9. **Nessel, R. J., DeKay, H. G., and Banker, G. S.,** Evaluation of polymeric materials. II. Screening of selected vinyls and acrylates as prolonged-action coatings, *J. Pharm. Sci.,* 53, 790, 1964.
10. **Bovey, F. A., Kolthoff, I. M., Medalia, A. I., and Meehan, E. J.,** *Emulsion Polymerization,* Interscience, New York, 1955, 1.
11. **Childs, E. S.,** *Am. Paint Coating J.,* p. 58, 1974; as cited in **Warson, H.,** *The Application of Synthetic Resin Emulsions,* Ernest Benn, London, 1972, 246.
12. **El-Aasser, M. S.,** Formation of polymer latexes by direct emulsification, 6th Annual Short Course, Advances in Emulsion Polymerization and Latex Technology, notes compiled by Poehlein, G., Center for Surface and Coating Research, Lehigh University, Bethlehem, Pa., 1975, 1.

13. **Lehmann, K. and Dreher, D.,** Use of aqueous synthetic polymer dispersions for coating of pharmaceutical dosage forms, *Drugs Made Ger.,* 16, 126, 1973.

14. **Bovey, F. A., Kolthoff, I. M., Medalia, A. I., and Meehan, E. J.,** *Emulsion Polymerization,* Interscience, New York, 1955, 15.

15. **Warson, H.,** *The Application of Synthetic Resin Emulsions,* Ernest Benn, London, 1972, 246.

16. Ethocel Standard, The Dow Chemical Form 170-229A-68, Plastic Coatings and Monomer Sales, The Dow Chemical Company, Midland, Mich., 1968.

17. **Luce, G. T.,** Cellulose acetate phthalate: a versatile enteric coating, *Pharm. Technol.,* 1, 27, 1977.

18. *New Standard Dictionary,* Funk and Wagnalls, New York, 1949; as cited in **Lehmann, K. and Dreher, D.,** *Drugs Made Ger.,* 16, 126, 1973.

19. **El-Aasser, M. S.,** Formation of polymer latexes by direct emulsification, 6th Annual Short Course, Advances in Emulsion Polymerization and Latex Technology, notes compiled by Poehlein, G., Center for Surface and Coating Research, Lehigh University, Bethlehem, Pa., 1975, 12.

20. **Bovey, F. A., Kolthoff, I. M., Medalia, A. I., and Meehan, E. J.,** *Emulsion Polymerization,* Interscience, New York, 1955, 10.

21. **El-Aasser, M. S. and Vanderhoff, J. W.,** Polymer Emulsification Process, U.S. Patent 4,177,177, 1979.

22. **Vanderhoff, J. W.,** Mechanism of film formation, *Br. Polym.,* 2, 161, 1970.

23. **Bondy, D.,** Emulsions and other aqueous media, in *The Science of Surface Coatings,* Chatfield, H. W., Ed., D Van Nostrand, Princeton, N.J., 1962, 393.

24. **Warson, H.,** *The Application of Synthetic Resin Emulsions,* Ernest Benn, London, 1972, 118.

25. **Harkins, W. D.,** A general theory of the mechanism of emulsion polymerization, *J. Am. Chem. Soc.,* 69, 217, 1951.

26. **Harkins, W. D.,** General theory of mechanism of emulsion polymerization. II, *J. Polym. Sci.,* 6, 217, 1951.

27. **Fitch, R. M.,** *Polymer Colloids,* Plenum Press, New York, 1971.

28. **Yeliseyeva, V. I.,** Effect of monomer polarity on latex polymerization, *Acta Chim. Acad. Sci. Hung.,* 71(4), 465, 1972.

29. **Vanderhoff, J. W., Bradford, E. B., Tarkowski, H. L., Shaffer, J. B., and Wiley, R. M.,** Inverse emulsion polymerization, *Adv. Chem.,* 26(34), 32, 1962.

30. **Ugelstad, J., El-Aasser, M. S., and Vanderhoff, J. W.,** Emulsion polymerization: initiation of polymerization in monomer droplets. *J. Polym. Sci. Polym. Lett. Ed.,* 11, 503, 1973.

31. **Medvedev, S. S.,** *Int. Symp. Macromolecular Chemistry,* Pergamon Press, New York, 1959, 147; as cited in **Warson, H.,** *The Application of Synthetic Resin Emulsions,* Ernest Benn, London, 1972, 118.

32. **Smith, W. V. and Ewart, R. H.,** *J. Phys. Chem.,* 16, 592, 1948; as cited in **Warson, H.,** *The Application of Synthetic Resin Emulsions,* Ernest Benn, London, 1972, 119.

33. **Aelony, D. and Wittcoff, H.,** U.S. Patent 2,899,397, 1959; as cited in **Warson, H.,** *The Application of Synthetic Resin Emulsions,* Ernest Benn, London, 1972, 246.

34. **Miller, A. L., Robinson, S. B., and Petro, A. J.,** U.S. Patent 3,022,250, 1962; as cited in **Warson, H.,** *The Application of Synthetic Resin Emulsions,* Ernest Benn, London, 1972, 246.

35. **Schoering, H., Witte, J., and Pampus, G.,** German Patent 2,013,359, 1971; *Chem. Abstr.,* 76, 26187J, 1972; as cited in **Warson, H.,** *The Application of Synthetic Resin Emulsions,* Ernest Benn, London, 1972, 246.

36. **Burke, C. W., Jr.,** U.S. Patent 3,652,482, 1972; *Chem. Abstr.* 77, 21211z, 1972; as cited in **Warson, H.,** *The Application of Synthetic Resin Emulsions,* Ernest Benn, London, 1972, 246.

37. **Cooper, W.,** U.S. Patent 3,009,891, 1961; as cited in **Warson, H.,** *The Application of Synthetic Resin Emulsions,* Ernest Benn, London, 1972, 246.

38. **Saunders, F. L. and Pelletier, R. R.,** U.S. Patent 3,642,676, 1972; *Chem. Abstr.,* 76, 142492e, 1972; as cited in **Warson, H.,** *The Application of Synthetic Resin Emulsions,* Ernest Benn, London, 1972, 246.

39. **Date, M. and Wada, M.,** Japanese Patent 73,06,619, 1973; *Chem. Abstr.,* 80, 38097b, 1974; as cited in **Warson, H.,** *The Application of Synthetic Resin Emulsions,* Ernest Benn, London, 1972, 246.

40. **Jud, P.,** British Patent 1,142,375, 1969; *Chem. Abstr.,* 70, 69355g, 1969; as cited in **Warson, H.,** *The Application of Synthetic Resin Emulsions,* Ernest Benn, London, 1972, 246.

41. **Schulman, H. J. and Cockbain, E. G.,** Molecular interaction at oil-water interfaces. I. Molecular complex formation and the stability of oil in water emulsions, *Trans. Faraday Soc.,* 36, 651, 1940.

42. **Ugelstad, J., Lervik, H., Gardinovacki, B., and Sund, E.,** Radical polymerization of vinyl chloride: kinetics and mechanism of bulk and emulsion polymerization, *Pure Appl. Chem.,* 26, 1921, 1971.

43. **Shinoda, K. and Friberg, S.,** Microemulsions: colloidal aspects, *Adv. Colloid Interface Sci.,* 4, 281, 1975.

44. **Vold, R. D. and Mital, K. L.,** The effect of lauryl alcohol on the stability of oil in water emulsions, *J. Colloid Interface Sci.,* 38, 451, 1972.

45. **Bovey, F. A., Kolthoff, I. M., Medalia, A. I., and Meehan, E. J.,** *Emulsion Polymerization,* Interscience, New York, 1955, 179.
46. **Vanderhoff, J. W., van der Hul, H. J., Tausk, R. J. M., and Overbeek, J. T. G.,** Well characterized monodisperse latexes, in *Clean Surfaces: Their Preparation and Characterization for Interfacial Studies,* Golffinger, G., Ed., Marcel Dekker, New York, 1970, 15.
47. **Severs, E. T.,** *Rheology of Polymers,* Rheinhold, New York, 1962, 64.
48. **Saunders, F. L.,** Rheological properties of monodisperse latex systems. I. Concentration dependence of relative viscosity, *J. Colloid Sci.,* 16, 13, 1961.
49. **Mooney, M.,** The viscosity of a concentrated suspension of spherical particles, *J. Colloid Sci.,* 6, 162, 1951.
50. **Brodnyan, F. F.,** The dependence of synthetic latex viscosity on particle size and size distribution, *Trans. Soc. Rheol.,* 12, 357, 1968.
51. **Saunders, F. L.,** Rheological properties of monodisperse latex systems: flow curves of thickened latexes, *J. Colloid Interface Sci.,* 23, 230, 1967.
52. **Woods, M. E. and Krieger, I. M.,** Rheological studies on dispersions of uniform colloidal spheres. I. Aqueous dispersions in steady shear flow, *J. Colloid Interface Sci.,* 34, 91, 1970
53. **Dillon, R. E., Mathenson, L. A., and Bradford, E. B.,** Sintering of synthetic latex particles, *J. Colloid Sci.,* 6, 108, 1951.
54. **Brown, G. L.,** Formation of films from polymer dispersions, *J. Polym. Sci.,* 22, 423, 1956.
55. **Voyutskii, S. S.,** Amendment to the papers by Bradford, Brown, and co-workers: concerning mechanism of film formation from high polymers, *J. Polym. Sci.,* 32, 528, 1958.
56. **Vanderhoff, J. W., Tarkowski, H. L., Jenkis, M. C., and Bradford, E. B.,** Theoretical considerations of the interfacial forces involved in coalescence of latex particles, *J. Macromol. Chem.,* 1, 361, 1966.
57. **Bradford, E. B. and Vanderhoff, J. W.,** Morphological changes in latex films, *J. Macromol. Chem.,* 1, 335, 1966.
58. **Bradford, E. B. and Vanderhoff, J. W.,** Additional studies of morphological changes in latex films, *J. Macromol. Chem.,* 6(4), 671, 1972.
59. **Vanderhoff, J. W.,** Mechanism of film formation of latices, *Br. Polym. J.,* 2, 166, 1970
60. **Vanderhoff, J. W., Bradford, E. B., and Carrington, W. K.,** The transport of water through latex films, *J. Polym. Sci.,* 41, 155, 1973.
61. **Miles, D. C. and Briston, J. H.,** *Polymer Technology,* Chemical Publishing, New York, 1965, 269.
62. **Walton, D. D.,** Cellulose for film manufacture, in *The Science of Polymer Films,* Vol. 1, Sweetening, O. J., Ed., Interscience, New York, 1968, 87.
63. **Arnold, L. K.,** *Introduction to Plastics,* Iowa State University Press, Iowa City, 1968, 68.
64. **Savage, A. B., Young, A. E., and Maasberg, A. T.,** Ethers, in *Cellulose and Cellulose Derivatives,* Vol. 5 (Part 2), 2nd ed., Ott, E., Spurlin, J. H., and Grafflin, M. W., Eds., Interscience, New York, 1963, 882.
65. **Savage, A. B., Young, A. E., and Maasberg, A. T.,** Ethers, in *Cellulose and Cellulose Derivatives,* Vol. 5 (Part 2), 2nd ed., Ott, E., Spurlin, J. H., and Grafflin, M. W., Eds., Interscience, New York, 1963, 914.
66. **Reidinger, F. J.,** Analytical chemistry of cellulose films, in *The Science of Polymer Films,* Vol. 1, Sweetening, O. J., Ed., Interscience, New York, 1968, 671.
67. Technical Data Publication No. ZFD-100, Eastman Kodak Chemical Products, Kingsport, Tenn., 1971.
68. **Luce, G. T.,** Cellulose acetate phthalate: a versatile enteric coating, *Pharm. Technol.,* 1, 27, 1977.
69. **Crawford, R. R. and Esmerian, O. K.,** Effect of plasticizers on some physical properties of cellulose acetate phthalate films, *J. Pharm. Sci.,* 60, 312, 1970.
70. **Amann, A. H., Lindstrom, R. E., and Swarbrick, J.,** Factors affecting water vapor transmission through polymer films applied to solid surfaces, *J. Pharm. Sci.,* 63, 931, 1974.
71. **Lee, V. H. and Robinson, J. R.,** Drug properties influencing the design of sustained or controlled release drug delivery systems, in *Sustained Systems,* Robinson, J. R., Ed., Marcel Dekker, New York, 1978.
72. **Hall, H. S. and Pondell, R. E.,** The Wurster process, in *Controlled Release Technologies: Method, Theory and Practice,* Vol. 2, Kydonieus, A. F., Ed., CRC Press, Boca Raton, Fla., 1980.
73. **Wheatley, T. A.,** *Sustained Release Phenylpropanolamine — Stability Considerations,* FMC Corporation, Princeton, N.J., 1984.
74. **Banker, G. S.,** Film coating theory and practice, *J. Pharm. Sci.,* 55(1), 1966.
75. **Banker, G. S. and Peck, G. E.,** The new, water-based colloidal dispersions, *J. Pharm. Tech.,* April 1981.
76. **Goodhart, F. W., Harris, M. R., Murphy, K. S., and Nesbitt, R. U.,** An evaluation of aqueous film-forming dispersions for controlled release, *Pharm. Technol.,* 8, 64, 1984.

Chapter 3

COLUMN FILM COATING AND PROCESSING

Harlan S. Hall and Jesse Wallace

TABLE OF CONTENTS

I. INTRODUCTION

Column coating is the use of modified fluidized-bed coating systems also known as air-suspension columns. These are known as columns because the coating chamber is actually a vertical column in which the material to be coated is circulated as it is fluidized by the process air. This should be distinguished from a simple fluidized bed in which the material is just barely fluidized but does not circulate. Simple fluidized beds can also be used for certain industrial coating processes by immersing heated parts into a fluidized bed of thermoplastic coating powder.[1-3] The thermoplastic powder melts and adheres to the heated parts, forming a continuous coating.

Figure 1 shows that there have been many variations in fluidized-bed coating systems; however, only two are commonly used to coat particles. The fluid-bed granulator is widely used to convert powders into larger granules and can be used to apply a coating in certain cases. There are two primary problems in using granulators as coaters: (1) granulators are designed to cause the particles to adhere to each other, which is usually undesirable in coating, and (2) the efficiency and uniformity of coating are poor compared to the Wurster process.

The Wurster column was invented in 1959 by Professor Dale E. Wurster.[4,5,7-10] It is designed to coat materials with little or no agglomeration of small particles into larger granules and characteristically deposits almost 100% of the coating uniformly onto the material being coated. The Wurster process is particularly well suited to coating, or encapsulating, small particles such as powders, crystals, granules, and beads. It is well suited for coating larger particles such as tablets and formed shapes as well. Examples will be given to demonstrate that this process is extremely versatile as it is capable of processing both a wide variety of particle sizes and shapes and also a tremendous range of coating materials.

II. COLUMN COATING SYSTEMS

As mentioned above, columns can be divided into granulators and Wurster columns. Granulators are characterized by a random or turbulent cycle of particles to be coated and be spray nozzles positioned at or near the top of the processing chamber. This upper location is useful when granulating because the smallest particles will rise through the spray and be wetted. It is undesirable when coating because the spray is counter-current to the process air flow and there is a tendency for the coating to spray dry and be lost rather than wet the particles and form a film.

Wurster columns overcome this difficulty by placing the spray nozzle at the base of the coating chamber and spraying the coating cocurrent with the airstream. The efficient and uniform deposition of coating is further enhanced by inducing ordered flow of the particles in the chamber so that the particles are presented in a controlled manner to the coating spray.

The Wurster coating chamber is usually cylindrical and has a perforated plate at the bottom (Figure 2). This plate is designed with larger holes in the center so that when air is drawn through the plate there is more air passing through the center than in the annular area. With a product in the coating chamber the air flow is adjusted so that the material in the annular area is just fluidized. Because more air is flowing in the center area, the product in that region ascends on the airstream forming a spout. As the spout removes product from the center, the fluidized mass in the annular area flows into the center and in turn ascends the spout. This action creates a circulating bed in which a constant flow of material is spouted up the center and descends in the annular region (Figure 3).

In many cases the ascending flow of the spout is separated from the descending flow at the outside by a tube mounted in the chamber. This tube is referred to as a partition because it divides, or partitions, the two flows. The partition is mounted so that a gap exists between

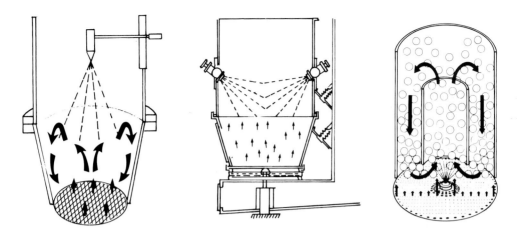

FIGURE 1.   Schematics of various fluidized-bed coating systems.

FIGURE 2.   Perforated plate at bottom of Wurster coating chamber.

the perforated plate and the bottom of the partition, which permits the product descending at the outside to reenter the spout. This gap is variable and is adjusted to regulate the flow into the center.

The spray nozzle is mounted at the center of the chamber near the plate. Various types of nozzles are used, most commonly two-fluid or hydraulic nozzles. The coating media is introduced through the nozzle and atomized. As the product flows under the partition and into the spout it passes through the spray pattern. In order to avoid agglomeration it is important that the particles separate from each other as they receive the spray. The particles receive a small amount of coating as they pass through the spray, which dries as the particles rise in the spout.

As the spout of product exits the top of the partition and enters the expansion chamber

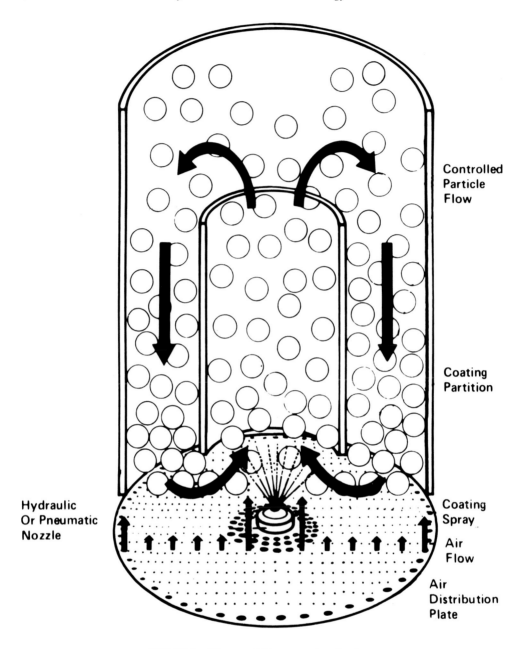

Controlled
Particle
Flow

Coating
Partition

Hydraulic
Or Pneumatic
Nozzle

Coating
Spray

Air
Flow

Air
Distribution
Plate

FIGURE 3.   Diagram of Wurster coating chamber.

(Figure 4) the flow expands and slows, allowing the product to fall out of the airstream and return to the product bed in the annular area of the coating chamber. In this region the product is fluidized and descends to the bottom to reenter the spray.

One entire cycle takes 5 to 10 sec to complete; thus the product receives a small increment of coating 6 to 12 times a minute. Because the coating is applied in small increments and dried after each, the product never becomes grossly wetted and the coating is gradually built up (Figure 5). The ordered flow causes the particles to circulate through the spray in a predictable manner so that all the particles have the same opportunity to receive coating. This results in deposition of highly uniform coatings from particle to particle.

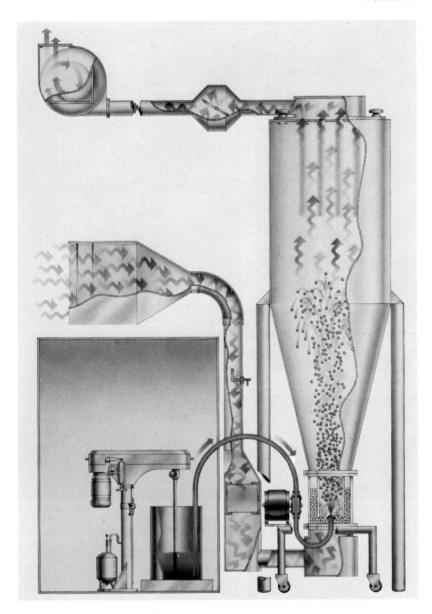

FIGURE 4. Expansion chamber of Wurster coating column.

III. UNIFORMITY OF COATING

The level of uniformity can be controlled by controlling the number of times the product circulates through the spray. Table 1 shows the increasing uniformity from particle to particle as a function of time, which reflects the number of cycles the particles have made. The uniformity was determined by spraying a coating containing a tracer chemical onto an inert core, then analyzing single cores for the presence of the tracer. The data show increasing uniformity (decreasing standard deviation) of particles within a batch of coated particles and also show the range of standard deviations for particles within multiple batches processed under similar conditions. Although this level of uniformity is not required for many applications, it can be readily achieved by adjusting the processing time.

Also of concern is the reproducibility of this uniformity between different pieces of

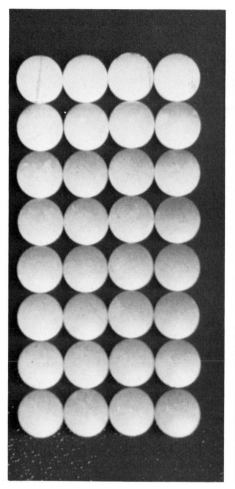 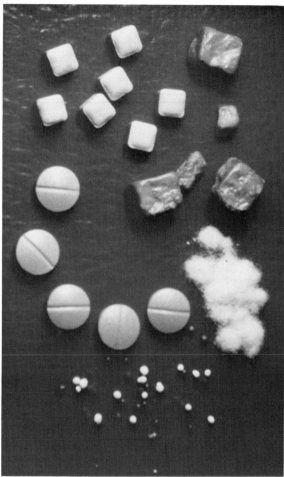

FIGURE 5.   Incremental coating of tablets by   FIGURE 6.   Variety of shapes that can be coated by Wurster
Wurster process.   process.

Table 1
COATING UNIFORMITY VS. TIME

Coating time (min)	Standard deviation (%) (1 batch)	Range (%) (multiple batches)
78	7.48	
86	6.15	5.7—6.6
100	4.80	
113	2.96	

equipment and different sizes of equipment. To test this, several trials were made using single-nozzle 18-in. units (3 ft³) and seven-nozzle 46-in. units (21 ft³). The particle-to-particle uniformity was measured for product from both units and also in the multiple-nozzle unit with a simulated malfunction of one nozzle.[11] Table 2 shows that the uniformity was good within each unit and also between units. In the case of the malfunction, the uniformity decreased slightly but was still quite good.

Table 2
COATING UNIFORMITY IN SCALE-UP

Coating unit	Standard deviation (%)
18-in. Wurster	7.2
46-in. Wurster	7.6
46-in. simulated-malfunction	9.4

IV. PROCESS CONTROL

A very important feature of coating columns is the excellent control and ease of instrumentation they permit. The size and shape of the particles largely determine the perforated plate that works best as well as the process air volume used in processing. The sensitivity of the material processed and the properties of the coating applied determine the processing temperatures which work best. The combination of processing air volume and temperature determines the maximum drying capacity for a given system and thus determines the maximum rate at which coating can be applied and dried. These variables are readily instrumented and controlled, making the process highly reproducible.

A potential limit to processing rate is the ability of the nozzle used to atomize the coating material.[12] It is important that the coating be well atomized and uniformly deposited onto the core particles. This is not a problem in small equipment because the spray rates fall well within the capacity of many commercially available two-fluid nozzles which typically do not atomize as well above 100 mℓ/min. In the larger equipment this rate is greatly exceeded, requiring the use of specially designed nozzles to deliver well-atomized coating at rates between 200 and 700 mℓ/min per nozzle. Above this range hydraulic nozzles work well with coatings in aqueous or organic solutions.

Another important consideration is selecting and setting the partition. Partitions may be simple straight-sided tubes or may be tapered to modify flow within the partition. Selection of the partition is based on flow properties of the particles being coated. Tapering the partition can result in better separation of particles as they pass through the spray from the nozzle, but may adversely affect material flow in the annular down-bed portion of the chamber.

The gap between the perforated plate at the bottom of the chamber and the base of the partition is very important. If this gap is set too wide (partition too high), too many particles will flow into the spray (coating zone) and they will not separate properly as they pass through the spray. This can result in agglomeration of the small particles as they are glued together by the coating material. If the gap is too small, too few particles will enter the coating zone. Coating material which fails to hit a particle will be spray-dried or will hit and dry onto the inside of the partition itself. Either extreme wastes coating material and may lead to other difficulties.

Although air flow, temperature, spray rate, nozzle selection, choice of plate and partition, and even the product being coated all interact to effect successful coating, these interact in a predictable manner and can be adjusted to give excellent results. Once the process has been optimized and the desirable process limits determined, the process is able to produce batch after batch with great consistency. Because of this highly reproducible nature of the process, it lends itself to automation. A process-control program has been implemented which controls the process within very close tolerances, shuts itself down in the event of mechanical failure, and documents each step of the process as it is performed.

V. CORE SELECTION

As mentioned above, column or fluid-bed processing is well suited for coating particulate

FIGURE 7.   Cross section of coated tablet showing even coating over irregular surface.

solids. It is ideal for coating (encapsulating) solids in the particle size range from 10 U.S. mesh (2000 μm = 2 mm) down to 140 U.S. mesh (100 μm = 0.1 mm), a range where few other processes are effective. It is possible to process particles down into the micron size range if some agglomeration of the particles is acceptable. Particles larger than 1 in. may also be coated, although alternate techniques are available for products of this size.

When small particles are processed, filters are used to prevent the loss of particles on the fluidizing airstream. Although various types of filters have been tried, filters which are self-cleaning without interrupting the process air flow or the spray pattern work best. When it is desirable to remove the dust from slightly larger particles the filters may be replaced by baffles designed to retain the larger particles while permitting the dust to be carried to external filters. When large particles are processed, filters are usually not required.

The ideal shape of particles to be coated is spherical with a smooth surface. Such particles will fluidize easily and require the least amount of coating because they have a minimum surface area. It is not required, or even common, that spherical particles be used. Most materials have a characteristic shape (salt crystals), a highly irregular shape (milled or ground particles), or a desired shape (extruded, molded, or punched shapes) (Figure 6). These irregular shapes can not only be coated by this technique, but will be coated evenly over the irregular surfaces (Figure 7).

Because large particles require more air to fluidize them than small particles, it is desirable to have a narrow size distribution of particles within a batch. Although starting with uniformly sized particles will yield the most uniform product, it is quite common that the starting material will include a range of particle sizes. A size range of 4:1 usually presents no special problems and wider size ranges may be accommodated, although particle-to-particle coating uniformity may suffer. Table 3 shows a typical size distribution for uncoated product and how the size increases to different levels as the product is coated.[12] Although a range of particle sizes is present in the starting material, it can be coated with little difficulty and only modest size increases.

In extreme cases this can be a problem, as the smallest particles represent dust mixed in with large particles. In these cases the dust will probably be carried off in the airstream to filters and the large particles will remain to become coated. If the small particles are allowed to remain in the coating chamber, they often adhere onto the larger particles, causing the

Table 3
PARTICLE SIZE VS. COATING LEVEL

		Coating Level (%)				
U.S. sieve	Microns	0.0	9.0	19.0	24.0	40.0
+ 30 mesh	>590	0.0	0.1	0.3	0.3	0.1
− 30/ + 40	590—420	0.0	0.1	0.1	0.1	0.6
− 40/ + 80	420—177	19.0	25.4	40.5	41.5	53.6
− 80/ + 120	177—125	42.7	40.9	40.2	38.2	28.1
− 120 mesh	<125	38.3	33.3	20.2	18.9	17.6

Table 4
REQUIRED COATING LEVEL (10 μm) vs. PARTICLE SIZE

U.S. mesh	Uncoated diameter (μm)	Surface area per gram (mm²)	Coated diameter (μm)	% Coating in product
5	4,000	1,157	4,020	1.18
10	2,000	2,312	2,020	2.34
18	1,000	4,610	1,020	4.49
35	500	9,535	520	8.75
60	250	18,490	270	16.70
120	125	36,917	145	30.20
200	74	63,004	94	45.10
325	44	107,018	64	62.00

formation of a few large "snowballs" which are easily removed from the single coated particles.

If possible, it is highly desirable to select the optimum particle size for the starting material. This is because surface area increases rapidly as particle size becomes smaller (Table 4). If a given coating thickness is required to achieve the desired result, this will be so whether the coating is applied over a large or small particle. To maintain the required film thickness over greater surface area, much more coating material is required. This leads to increases in both time and materials cost. In addition, this coating thickness greatly increases the relative size of the smaller particles compared to the larger particles.

Very sensitive particles can be processed by this method. Material which is sensitive to heat and moisture can be successfully coated by using the high drying capacity of this process along with modest process temperatures and carefully adjusted spray rates. Examples of this are agricultural seeds (Figure 8), certain of which rapidly lose viability when exposed to high humidity or to temperatures of 110° F or above. Alternatively, materials which are highly hygroscopic may be coated with water-based coatings without being adversely affected by the water used. Examples are the coating of citric acid with a water-soluble coating and the coating of calcium chloride with a water-based coating. This can be accomplished even without dehumidifying the process air because the process air can be heated to a low relative humidity and the contact time of the liquid droplets before they dry is short.

Certain pharmaceutical products are sensitive to moisture, heat, or both. The use of aqueous coatings on sensitive cores requires that the moisture be removed very rapidly before it can adversely interact with the core. If the product is heat sensitive, it is necessary to achieve this rapid drying without use of excessive heat. The excellent drying characteristics of the Wurster process make it well suited to meet this requirement. Data published in 1973 established that such materials can be coated with water-based coatings without harming the

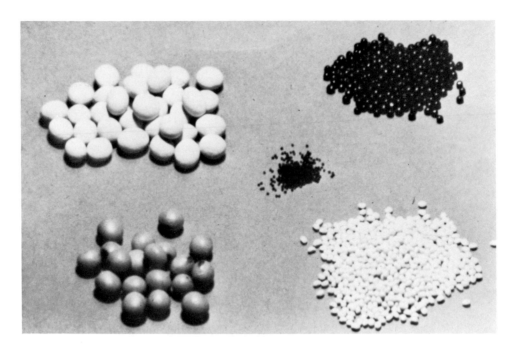

FIGURE 8.   Agricultural seeds that can be successfully coated, although sensitive to heat and moisture.

Table 5
DECOMPOSITION OF VITAMIN C (mg
DEHYDROASCORBIC ACID PER 50 mg
VITAMIN C)

Tablets	10—14 days (mg)	Accelerated storage (mg)
Uncoated	0.19	0.12
Coated	0.15	0.17

product.[13] Since that time a number of pharmaceutically approved aqueous coating materials have been made available commercially, which has contributed greatly to the increased use of aqueous coatings. Tables 5 and 6 show the stability of two water- and heat-sensitive products coated with aqueous systems.

VI. COATING SELECTION

A. Solvents

The great variety of coating materials which can be utilized in column coating tremendously enhances the versatility of fluid-bed processing. The solvent-based film-coating formulations commonly used in conventional and vented coating pans also work well in fluid-bed processing. When working with organic solvents one must be concerned with the potential exposure of workers to solvents, the flammability of the solvents, and the residual solvent in the product. Fluid-bed processing can be used to address each of these concerns. Worker exposure is minimized because all processing is performed in a closed system. Flammability concerns are minimized by explosion venting of equipment and by dilution of solvent concentrations to less than the lower explosive limit (LEL).[14,15] Residual solvent problems are controlled by utilizing the drying capacity of the equipment to reduce traces of solvent to extremely low residuals.

Table 6
DECOMPOSITON OF ASPIRIN (% SALICYLIC ACID)

Tablets	Process air temperature (°F)	10—14 days (%)	Accelerated storage (%)
Uncoated		0.09	0.23
Coated, A	130	0.14	0.48
Coated, B	115	0.12	0.18
Coated, C	110	0.12	0.84

B. Aqueous

Many other less conventional coatings are also readily used. The excellent drying capacity of the equipment makes it ideally suited for use with aqueous systems including water-soluble polymers and gums, aqueous latex and pseudolatex systems, and aqueous suspensions. Latex and aqueous suspensions are of particular interest because they may be applied at higher solid contents than in many solution systems. This results in favorable processing times even though water is used as the vehicle.[16]

C. Hot Melts

In addition, coatings which can be melted are applied in a molten condition and solidified onto the core. A wide variety of fats, waxes, surfactants, and emulsifiers, and even some polymers, may be applied as hot melts. This can be very economical because the coating is sprayed at 100% solids, resulting in very rapid build-up of high coating levels. The very short processing times typical of hot-melt systems usually result in a need to use a slightly heavier coating than other systems. The reason is simply that sufficient processing time must be used to permit build-up of a uniform coating (see Section III).

A partial list of coating materials which have been used in coating columns is found in Table 7. This list includes examples of each of the above as well as other materials, but should not be considered exhaustive.

VII. APPLICATIONS

It is impossible to give the details of many of the products processed in fluid-bed equipment because the information is restricted by confidential agreements. The following are examples chosen because the information is available and they illustrate different approaches to controlled release.

A. Taste Masking

Many products of interest have an objectionable taste, making regular oral dosing of the product a problem. In both humans and domestic animals acceptability of medicines can be greatly improved by masking the objectionable taste before incorporating the drug into a palatable dosage form. The product may be a pharmaceutical dosage form, an animal feed ingredient, or even a toxin. An example is the rodenticide Warfarin®.[17,18] In the early 1970s it became apparent that mice and rats were able to avoid toxic baits used as rodenticides. A test was devised by the Environmental Protection Agency in which a toxic bait and a prescribed feed were fed free-choice to mice. The test required that a minimum percentage of the food eaten in a free-choice situation be from the toxic bait or it is considered ineffective. Figure 9 shows the improved acceptability of the toxic bait after encapsulation of the rodenticide.

Also of concern in taste-masking applications is the bioavailability of the core. If the taste

Table 7
COATING MATERIALS USED IN
COATING COLUMNS

Aqueous solutions, solvent solutions,
latices, emulsions, dispersions, and melts

Acrylics
Carboset®
Carbowax®
Casein
Cellulosics
    Cellulose acetate (CA)
    Cellulose acetate butyrate (CAB)
    Cellulose acetate phthalate (CAP)
    Ethyl cellulose (EC)
    Hydroxyethyl cellulose (HEC)
    Hydroxypropyl cellulose (HPC)
    Hydroxypropyl methyl cellulose (HPMC)
    Hydroxypropyl methyl cellulose phthalate (HPMCP)
Chitosan
Chlorinated rubber
Clays
Coating butter
Corn syrup
Dextrins
Enterics
Eudragits
Ethylene vinyl acetate (EVA)
Fats
Gelatin
Glycerides
Vegetable gums (arabic, carrageenan)
Halocarbons
Hydrocarbons
Hydrocarbon resins
Hydrogenated oils
Kynar®
Microcrystalline wax
Milk solids
Molasses
Nylon
Opadry®
Paraffin wax
Phenolics
Pluronics
Polyethylene
Polyglutamic acid
Polylactides
Polyvinyl acetate (PVA)
Polyvinyl alcohol (PVAL)
Polyvinyl chloride (PVC)
Polystyrene
Polyvinyl acetate phthalate (PVAP)
Polyvinylidene chloride (PVDC)
Polyvinylpyrrolidone (PVP)
Proteins
Rubber, synthetic
Shellac
Starches
Stearates
Stearines

Table 7 (continued)
COATING MATERIALS USED IN COATING COLUMNS

**Aqueous solutions, solvent solutions,
latices, emulsions, dispersions, and melts**

Sucrose
Surfactants
Teflon®
Vermiculite
Waxes, vegetable
Waxes, mineral
Zein

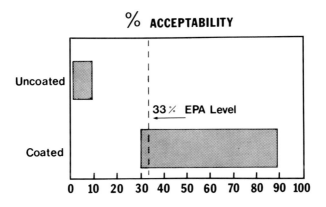

FIGURE 9.   Improved acceptability of toxic bait after encapsulation.

is well masked but the active ingredient is not released, there will be little benefit in the product. Figure 10 shows the plasma levels achieved for a veterinary product, the taste of which has been masked. The bioactive material in crystalline form is coated, mixed with feed ingredients, and compacted before dosing. The animals being treated readily eat the end product, which was designed to release its contents soon after swallowing. Figure 10 indicates that excellent bioavailability has been achieved, with plasma level curves for the coated and uncoated product being nearly identical.[19]

B. Sustained Release

Usually this type of controlled release is achieved by diffusion of the active ingredient through a semipermeable membrane or film, deposited as a coating around the core. By varying the thickness and composition of this film the release rate can be adjusted over a wide range. Achieving this control implies that the films or membranes must be deposited in a highly reproducible and uniform manner (see Section III). An example of a commercial sustained release product is Isoclor®, marketed by Fisons. This product is designed to release its contents over a 12-hr period. This is achieved by depositing a semipermeable membrane around small beads which contain the active ingredients.

Another, and more unusual, sustained release product is a cough syrup manufactured by Pennwalt Corp.[20] This product utilizes an ion-exchange resin complex with dextromethorphan. The core consists of small particles of ion-exchange resin complexed with the drug and specially treated. The treated particles are then coated with a semipermeable coating which controls the transport of ions into and out of the complex, thus controlling the release

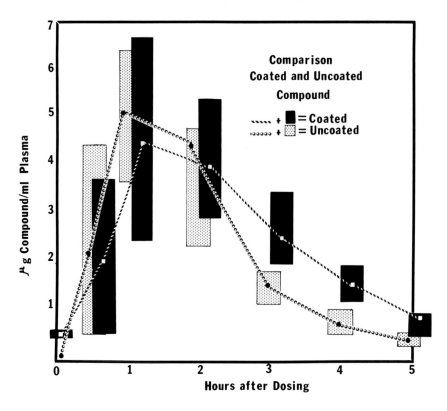

FIGURE 10.   Comparison of plasma levels for coated and uncoated product.

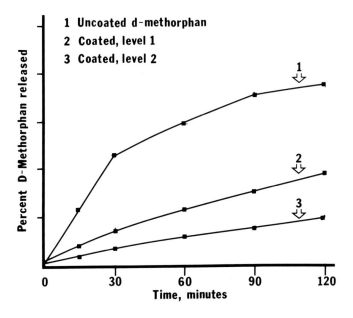

FIGURE 11.   Sustained release achieved by different levels of coating. (From Raghunathan, Y., Amsel, F., Hinsvark, O., and Bryant, W., *J. Pharm. Sci.*, 70, 379, 1981. Reproduced with permission of the copyright owner, the American Pharmaceutical Association.)

of the drug. Figure 11 illustrates the sustained release achieved by the resin complex alone and with different levels of coating applied.

C. Delayed Release

This can be achieved by choosing coatings which are resistant to their usage environment until it is changed by a selected release mechanism. A common example is enteric coating of pharmaceuticals which are irritating to the stomach. These polymers are designed to be soluble above a certain pH, but insoluble below it. Delayed release can also be achieved by using coatings which are insoluble but melt at a certain temperature.

A more unusual delayed release product is Fintrol Lampricide® which is used to control sea lampreys. The toxin in Fintrol® is antimycin A which is toxic to gill-breathing species, including fish and lampreys. In order to poison the lamprey without harming fish, it is necessary to deliver the toxin into the mud where the immature lamprey live. This is accomplished with an erodible coating which is applied over the toxin. The topcoat prevents release of the toxin as the particles fall through the water, but the toxin is released into the mud only after the particles reach the bottom. Antimycin has a very short half-life and so must be released in a short time to be effective. Because the toxin degrades so quickly there is no residual toxic effect and no migration of the toxin into other areas.

Another variation on this is Fintrol 5® which is designed to treat only the top 5 ft of a body of water. This product is designed with the toxin in the erodible coating so that it is released as the particle falls through the water until the coating is gone.[21]

VIII. SUMMARY

The above overview attempts to convey the great versatility of column coating for applying a wide variety of coatings onto a wide variety of particles. Clearly, many of the details of how to formulate the coatings and what pitfalls await the inexperienced are beyond the scope of this book. Although many of the specific examples cannot be discussed due to confidentiality, much of what has been learned over the years is available through those experienced in using coating columns. It is my hope that this discussion will stimulate the reader to explore this technology as a tool for developing new products.

REFERENCES

1. **Gilbert, L. O.,** Organic Coating Using the Fluidized Bed Technique, Tech. Rep. #62-3183, Rock Island Arsenal Laboratory, Rock Island, Ill., Spetember 20, 1962.
2. **Chen, W. H. and Gutfinger, C.,** An approximate theory of fluidized bed coating, Symp. Fluidization and Fluid Technology, Part I, 61st Annu. Meet., Los Angeles, December 1968.
3. **Anon.,** Fluidized beds take on new life, *Chem. Eng. News,* p. 46, December 1970.
4. **Wurster, D. E.,** Means for Applying Coatings to Tablets or Like, U.S. Patent 2,799,241, 1957.
5. **Wurster, D. E.,** Granulating and Coating Process for Uniform Granules, U.S. Patent 3,089,824, 1963.
6. **Lindloff, J. A.,** Apparatus for Coating Particles in Fluidized Bed, U.S. Patent 3,177,027, 1964.
7. **Wurster, D. E.,** Apparatus for Encapsulation of Discrete Particles, U.S. Patent 3,196,827, 1965.
8. **Wurster, D. E.,** Process for Preparing Agglomerates, U.S. Patent 3,207,824, 1965.
9. **Wurster, D. E.,** Particle Coating Apparatus, U.S. Patent 3,241,520, 1966.
10. **Wurster, D. E.,** Particle Coating Apparatus, U.S. Patent 3,253,944, 1966.
11. **Hall, H.,** Uniformity of coatings applied in 46'' Wurster columns, The Coating Place technical paper, December 1978.
12. **Hall, H. and Pondell, R.,** Performance of two fluid nozzles, the Coating Place technical paper, January 1978.

13. **Hall, H. and Hinkes, T. M.,** Air suspension encapsulation of moisture-sensitive particles using aqueous systems, in *Microencapsulation: Processes and Applications,* Vandegaer, J. E., Ed., Plenum Press, New York, 1974, 145; as presented at Symp. Microencapsulation Processes and Applications, August 27 to 31, 1973.

14. **Pondell, R.,** Lower explosive limit for flammable solvents, The Coating Place tech. paper, February 1984.

15. **Bartknecht, W.,** Explosion Protection Measures on Fluidized Bed Spray Granulators and Fluidized Bed Driers, Central Safety Service, Ciba-Geigy, Basel, August 1975.

16. **Hall, H. S., Lillie, K. D., and Pondell, R.,** Comparison — aqueous vs. solvent based ethylcellulose films, The Coating Place technical paper, March 1980.

17. **Hall, H. S. and Hinkes, T. M.,** The Wurster process for controlling pesticides, presented at Controlled Release Pesticide Symp., University of Akron, Akron, Ohio, September 13 to 15, 1976.

18. **Abrams, J. and Hinkes, T. M.,** Acceptability and performance of encapsulated Warfarin®, *Pest Control,* p. 14, May 1974.

19. **Hinkes, T. M.,** Encapsulation of small particles, presented at 19th Annu. National Industrial Pharmaceutical Research Conf., Lake Delton, Wis., June 6 to 10, 1977.

20. **Raghunathan, Y., Amsel, F., Hinsvark, O., and Bryant, W.,** A new sustained release drug delivery system. I and II, *J. Pharm. Sci.,* 70, 379, 1981.

21. **Gilderhus, P. A.,** Efficacy of Antimycin for Control of Larval Sea Lampreys *(Petromyzon marinus)* in Lentic Habitats, Techn. Rep. #34, Great Lakes Fishery Commission, Ann Arbor, Mich., May 1979.

Chapter 4

APPLICATION OF COATINGS USING A SIDE-VENTED PAN

Harry Thacker, Eric Forster, and Jesse Wallace

TABLE OF CONTENTS

I. INTRODUCTION

The advent of the side-vented coating pan around 1968 introduced a new concept to the application of film coatings to tablets. Prior to this, those companies which carried out this process used either a suitably modified traditional coating pan or an air-suspension coater (Wurster column). These processes invariably used organic solvents with the film-forming polymers in solution.

The first side-vented pan was developed by Eli Lilly. Using this type of equipment for film coating greatly improved the process. By passing air through the tablet bed instead of only over the surface, the high drying (evaporative) capacity of the column was combined with the gentler action of the conventional coating pan, giving greater flexibility to the process.

The introduction of this type of pan led to increased use of the film-coating process within the pharmaceutical industry. This in turn led to the increased use of organic solvents and attention was then focused on the problems associated with their use and disposal. Apart from their high cost, many of these solvents are toxic and/or inflammable and, in the case of the inflammable solvents, expensive flameproof equipment is required to ensure safety.

The obvious solution was to use aqueous-based coatings rather than those based on organic solvents. This had proved difficult in conventional pans due to the high latent heat of evaporation of water (approximately three times that of the commonly used organic solvents), but it was more practical in the air-suspension column with its higher evaporative capacity. The introduction of the side-vented pan, combining the advantages of both types of equipment, has been one of the major factors leading to a substantial increase in the use of aqueous-based coatings.

Over the past 15 years the concept of the original side-vented pan, known as the Accelacota, has been extensively developed. It can now be used for all types of film coating and for sugar coating. Its greatest advantage lies in the fact that, when used in conjunction with a carefully designed coating system, it can give a very evenly distributed coating. This is desirable because less material and time are required to obtain a complete cover of the product being coated. This is particularly important in the case of sustained release products where the dissolution profile of the drug can be adversely affected by unevenness of the coating.

Because it is suitable for both sugar and film coating it has an additional use in field of sustained release technology. It is ideal for products having an active material in the core, which are then film coated with, for example, an enteric coating followed by a sugar coating containing a further quantity of the active material. Using the Accelacota it is possible to carry out the film coating and the sugar coating without removing the cores from the pan; the complete process can be automated.

II. DESCRIPTION OF THE METHOD OF OPERATION

As has been stated, the main difference between a side-vented pan and a conventional coating pan is that in the side-vented pan the drying air flows through the bed of tablets instead of across the surface of the bed. Different versions of the side-vented pan have different air-flow patterns; however, the Accelacota will be used to describe the principle of the method of operation, with comments on some of the variations to this design.

The air flow through the pan is shown in Figure 1. The drying air enters the pan through the perforated wall. It is introduced by means of a plenum (Figure 1A) which is sealed against the rotating perforated wall. The drying air then flows across the pan along with the droplets of coating material from the sprays. The air flow directs the droplets onto the surface of the bed of tablets or granules. The air then passes through the bed together with the

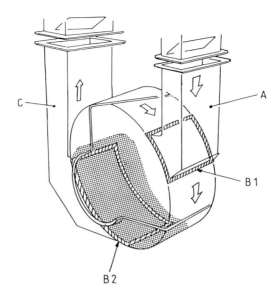

FIGURE 1.   Air flow through a side-vented pan.

solvent vapor which has evaporated from the droplets of coating solution. The air and vapor then leave the pan via the perforations behind the bed and enter a second plenum (Figure 1C), from which they are drawn out of the system and exhausted or recirculated according to the system being used. Provided the seals (Figure 1B1 and B2) on the plena (Figure 1A and C) are maintained in good condition, all the drying air will pass through the bed of tablets or granules.

Separate fans are used to convey the heated air into the pan and to exhaust the air/vapor mixture from the pan. The flow of air into and out of the pan is regulated by means of dampers to give a slightly negative pressure in the pan. This prevents the loss of any heated air, solvent vapor, or dust to the coating room.

The variations of this basic method consist mainly of different means of introducing and exhausting the air. One alternative method is to introduce the air along the axis of rotation, either via the mouth of the pan (which restricts access) or from the back of the pan. This method is believed to make the seals between the ducts and the pan simpler, but it causes the flow of air initially to be at right angles to the flow of droplets from the sprays. This results in a number of undesirable conditions. The spray pattern is disrupted, which in turn results in uneven coating and losses of coating material because part of the spray is carried onto the surface of the pan instead of the tablets. It also results in a variation in drying conditions across the bed.

With the other method the air flows in the opposite direction. It is introduced behind the tablet bed and leaves either along the axis of rotation or via the perforated surface opposite the bed. The two advantages of this method are that

1.   The introduction of air from behind the bed (provided it is at a sufficient velocity) reduces the pressure of the bed on the surface of the pan, thereby reducing any abrasion which might occur from tablets rubbing against the pan surface.
2.   The drying air, flowing countercurrent to the spray, results in more efficient heat transfer between the air and the spray.

There are, however, fallacies in these claims:

1. Under normal conditions the tablets lying on the surface of the drum move with the drum; therefore in normal operation there is virtually no abrasion between the tablets or granules and the drum. The introduction of air is likely to induce movement between the bed and the drum rather than prevent it. Any abrasion that does occur within the bed is more likely to result from tablet-to-tablet contact.
2. The air flowing countercurrent to the spray is likely to carry some of the spray away from the tablet bed. Any savings in heat transfer are likely to be more than outweighed in loss of materials. If the countercurrent air flow is increased such that tablet-to-tablet contact is reduced, the detrimental effect on the distribution of the spray would be considerable.

Another variation of the system uses no seals between the plenum and the rotating pan. In this case, four plena and associated ducting are welded to the pan to prevent the possibility of drying air bypassing the bed. However, the maintenance of the rotary seals connecting the fixed ducting to the rotating ducting presents an equally difficult maintenance problem. While the ducts welded to the pan can be cleaned, verification of the degree of cleanliness can be a problem. The various types of side-vented pans must therefore be judged on their merits for the particular coating function which they are required to perform.

III. THE FILM-COATING PROCESS

The successful application of a film coating to tablets or granules depends on a number of conflicting factors. Many are interrelated, making it particularly difficult to visualize the overall concept of the process.

Film coating differs considerably from sugar coating in that little or no tablet-to-tablet transfer of the coating takes place. For most of the commonly used film-coating materials, each drop of the film-coating solution only coats the spot on the surface of the tablet or granule on which it lands. It is therefore necessary to expose the whole surface of every tablet or granule to the spray before complete coverage is obtained.

This is not always true for some of the more recently introduced film-coating materials, as some of these films have the ability to spread and then coalesce. However, the speed and efficiency of the process is still governed by the rate at which new tablet or granule surfaces can be exposed to the spray.

The ideal film-coating process would deposit a given amount of film former, as evenly as possible and in the shortest possible time, onto a batch of tablets or granules. To achieve this the coating system would have to dry the coating at every rate at which it was applied and evenly cover the surface of the tablets or granules at every rate at which the new surface was exposed.

The three major functions governing the process, therefore, are (1) the mixing ability of the coating pan, (2) the spray system, and (3) the drying ability (i.e., evaporative capacity) of the system. Based on these factors two conclusions can be drawn:

1. There will be a minimum amount of solvent which can act as a carrier to distribute the coating evenly over the surface of the tablets or granules.
2. There will be a minimum time over which the coating can be applied.

IV. MIXING

The efficient coating of tablets or similar materials involves applying a complete and even cover of coating material in the shortest possible time. This should be achieved without

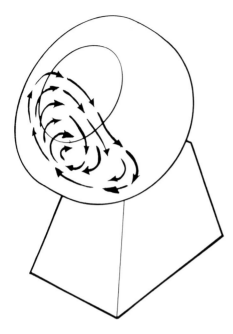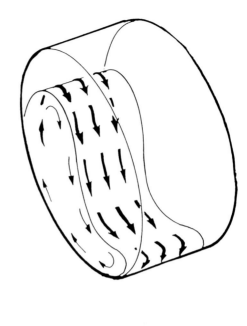

FIGURE 2.   Tablet movement in a conventional coating pan, showing the formation of a vortex.

FIGURE 3.   Tablet movement in a cylindrical pan.

unnecessary waste of coating materials or energy. The most important factor involved in achieving this result is the ability to expose the uncoated surface of the tablets to the spray as frequently as possible, i.e., mixing. The degree of mixing must in turn be related to the rate at which the coating fluid can be sprayed. There must also be a relationship between the degree of mixing and the rate at which the coating fluid can be sprayed. The mixing must also be carried out without damage to the material being coated or to any coating already applied.

Assuming the material to be coated is a batch of tablets which are suitably resistant to wear, the desired rate of exposing new surface to the sprays can often be achieved simply by running the pan at a sufficiently high speed. However, this situation rarely works in practice because tablets are frequently friable. It has therefore been necessary to devise a means of efficient mixing which is suitable for less wear-resistant tablets.

The conventional coating pan which produces a vortex within the tablet bed is probably the least efficient mixing method (Figure 2). Some tablets pass through the spray frequently while others pass through only occasionally. It is possible to film-coat tablets in a conventional pan, but it is necessary to spray slowly in order to limit the unevenness of the coating. The traditional pan can only give an even and consistent coating thickness with coating materials which have the ability to spread or to transfer from one tablet to another.

A better method for coating tablets evenly is to use a cylindrical pan, i.e., a true cylinder with flat sides (Figure 3). With this pan design tablets tend to follow a very consistent path with little or no movement from one side of the pan to the other. Successful coating, i.e., even distribution across the width of the bed, is difficult to achieve. If a nozzle producing a fan-type pattern is used, the quantity of material delivered in the center is normally much greater than at the edges. The tablets in the center of the pan consequently receive more coating. It is therefore advisable to introduce some form of side-to-side movement of the tablets within the pan. This can be achieved simply by using conical sides instead of flat sides for the pan (Figure 4). Even with this arrangement, however, mixing is still poor at

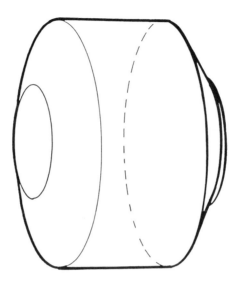

FIGURE 4.   A cylindrical coating pan with conical ends.

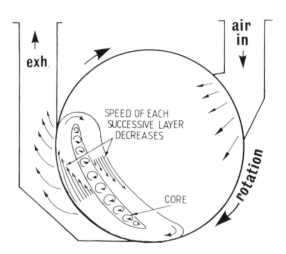

FIGURE 5.   Formation of a core within the tablet bed. (There is little intermixing between the core and the remainder of the bed.)

low pan speeds and it is often unsuitable for friable tablets. To achieve good mixing at low pan speeds some form of baffle arrangement is necessary.

In designing the mixing baffles used in the Accelacota, a study was made of tablet movement within the bed. Flow patterns were established by inserting a probe into the bed at various points to determine the direction of movement at different depths. Although the use of a probe disrupted the flow pattern, it was possible to establish general indications of tablet behavior. The results indicated that tablets on the surface of the bed and against the pan moved most rapidly, as would be expected. The speed of successive layers gradually decreased as they approached the core (Figure 5). These results were confirmed in larger beds (350 and 650 kg) using a large-diameter (150 mm) probe of a different type. This probe was fitted with a transparent calibrated window at the end inserted into the bed and was used in conjunction with multicolored tablets. It showed that tablets which entered the

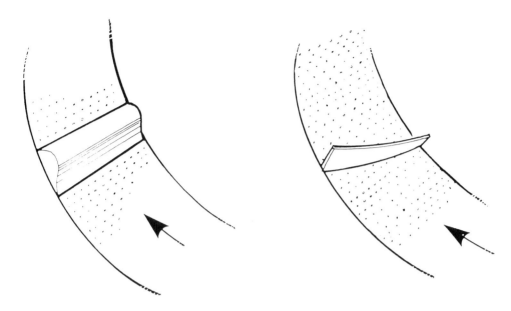

FIGURE 6. Simple mixing baffles.

core tended to stay within it for a considerable time before leaving and consequently received less coating. It was found that the core was less evident in shallower beds but became more pronounced with tablets of small diameter. It was also noted from these tests that the passage of air through the bed altered the tablet flow but did not change its general pattern.

This method was then used to assess the effectiveness of different types of baffles. It was observed that any mixing baffle which consists of a blade or series of blades attached to the cylindrical part of the pan, either parallel to the axis of rotation or at an angle to it (Figure 6), produced mixing only when entering or leaving the tablet bed, i.e., when the tablet movement changed direction. For the remainder of the time that the blades were passing through the bed they simply acted as driving plates, causing tablets to move through the bed at the speed of the pan.

A baffle arrangement which has been used successfully in the Accelacota consists of a series of curved blades mounted on the conical sides of the pan (Figure 7). They are designed to produce minimum disturbance of the bed at entry and exit to avoid damaging tablets. They are situated such that tablets can pass above and below them and no core of tablets is formed. As they pass through the bed, the tablets immediately behind them are not subjected to the weight of the layers of tablets above them and this causes them to move at a different speed than the remainder of the bed. As each baffle leaves the bed the tablets immediately behind it are deposited in a different position relative to the remainder of the tablet bed. This action, together with the differential speeds within the tablet bed caused by the baffles, gives the required degree of mixing. The advantage of this type of baffle in a side-vented pan has been demonstrated by Leaver et al.[1]

In the larger pans holding 350 kg or more, even this type of baffle was not sufficient to eliminate the core. A different baffle arrangement was designed. This consisted of large divergent tubes attached to each side of the pan and at 180° to each other (Figure 8). These act by removing tablets from the middle and lower layers of one side of the pan and depositing them on the surface of the bed at the opposite side.

The effectiveness of these baffles has been demonstrated by dividing the pan into two sections at right angles to its axis of rotation. The two sections were then loaded with tablets of the same size and shape but of different colors. The division was then removed and samples taken from various points within the bed. Photographs of the bed surface were taken

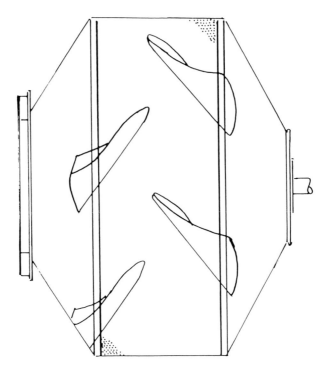

FIGURE 7.   A more effective type of mixing baffle.

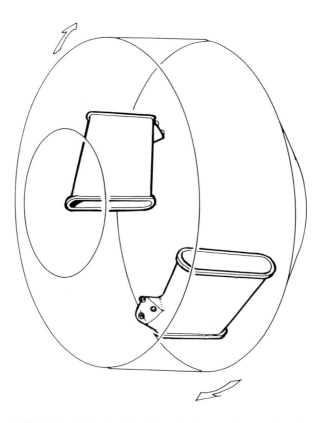

FIGURE 8.   Tubular baffles designed to disperse the core shown in
Figure 5.

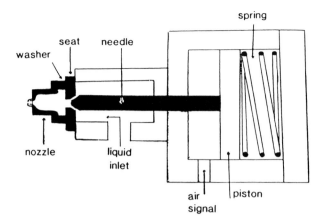

FIGURE 9. Operating principles of an airless spray gun.

after every two revolutions of the pan. They showed that after six revolutions complete mixing had occurred and that the distribution of the colors remained constant with further revolutions of the pan. Mixing is achieved irrespective of pan speed, therefore, very low speeds, e.g., 2 rpm, can be used for friable tablets. It has also been found that this type of mixing arrangement was able to evenly distribute a small quantity of tablets of different sizes, shapes, and densities. It may therefore be capable of eliminating the differential coating which normally takes place when tablets of different sizes and shapes are included in a single batch for experimental purposes.

V. SPRAY SYSTEMS

Mixing is one of the keys to the efficient, even coating required for sustained release. However, there is no advantage to be gained from good mixing unless the spray system is capable of evenly depositing the coating solution on the exposed surface.

The object of the spray system is to atomize the coating solution into droplets of a suitable size. The best size for a particular operation will depend upon the properties of the film former and coating solution. It will also depend upon the sensitivity of the material being coated to the coating solution and on the drying capacity of the system. The mean droplet diameter and droplet size range are to a large extent governed by the particular spray system being used, although they tend to fall in the range of 10 to 50 μm.

For efficient coating it is important to keep the droplet size range as narrow as possible. If the range is too wide it will be impossible to set drying conditions to suit all sizes. Either the smaller droplets will become too dry to adhere to the tablet surface or the largest droplets will cause overwetting of the tablets. The drying of the small droplets wastes material and energy and the overwetting of the tablets results in physical or chemical damage and a deterioration of the quality of the coating.

Two main types of spray guns are available: airless spray guns (Figure 9), which operate by forcing the coating fluid through a small nozzle at high pressure, and air-operated or conventional spray guns (Figure 10), which use compressed air to atomize the liquid. Airless sprays have been shown to give the best results with high liquid flows, as under these conditions they give a more consistent droplet size. If low liquid flows are required, either the nozzle size must be reduced such that it becomes difficult to operate or the liquid pressure must be reduced, leading to poor atomization. Consequently, airless sprays are most suitable for coatings using organic solvents where the low latent heat of evaporation allows the solution to be applied at a high rate. For aqueous coatings or other situations in which it is

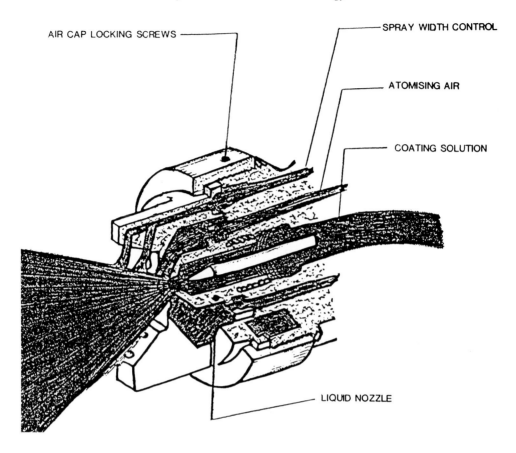

AIR CAP LOCKING SCREWS

SPRAY WIDTH CONTROL

ATOMISING AIR

COATING SOLUTION

LIQUID NOZZLE

FIGURE 10.   Operating principle of an air atomized spray gun.

necessary to use a lower liquid flow rate (e.g., cases where the coating goes through a tacky phase during drying) air-atomized sprays give the best results. The atomizing air is able to give good atomization at low liquid flow rates but the atomization deteriorates once the flow rate per gun has exceeded a certain value.

Airless spray systems normally employ piston pumps to generate the high liquid pressure required. These can be driven by air motors, which is convenient where flammable solvents are being used. However, the seals on these pumps can sometimes prove difficult to maintain when the coating solution contains abrasive materials such as pigments based on iron oxide or titanium dioxide.

The supply of coating solution at low pressure to the air-atomized spray guns can be achieved by a variety of means. Low-pressure piston pumps are frequently used; sometimes gear pumps are also used but abrasive components in the coating solutions can cause problems with both types of pump. Peristaltic pumps have proved very useful as they are easy to clean and maintain. These are positive displacement pumps, i.e., liquid flow can be accurately maintained. It is important not to use these pumps for coating solutions containing organic solvents which may leach the plasticizer out of the rubber. Although the rubber tubes would then have to be replaced more often, the more serious problem is that the plasticizer would be deposited in the coating.

VI. DRYING (EVAPORATIVE) CAPACITY

In a system with good mixing where the tablet surface is regularly exposed to the spray with good distribution across the tablet bed and satisfactory droplet size, the one remaining requirement for efficient coating is the drying system. A well-designed coating plant must have a drying system which is capable of drying the solvent from the coating solution at whatever rate it can be applied. There is little point in having good mixing and spraying if the spray rate has to be reduced because of overwetting. It is therefore advisable to have some slight excess capacity built into the drying system to ensure that it is adequate in all conditions. The additional capital cost is small compared to the financial value of improved efficiency or the cost of upgrading a system with insufficient capacity.

Overcapacity in a drying system should not result in increased running costs. If the air flow and temperature are adjusted to give the required amount of drying, operating costs will be similar irrespective of the total capacity of the system.

Much comment has been made regarding the thermal efficiency of coating systems, often to the exclusion of the overall efficiency of the total coating system. Thermal efficiency should never be considered the sole criterion of efficient operation. This may be illustrated by two cases. In the first, the requirement is to coat only one batch per day. Provided the tablets can withstand the abrasion of rolling for long periods without the protection of a coating, the operation can proceed very slowly. Room temperature air can be used for the drying and the cost of evaporation kept to a minimum. This is the most efficient operation for these circumstances and can give an excellent quality coating even with relatively poor mixing and poor spray conditions. In a second case, there is a shortage of capacity. Coating an extra batch in a working day will save the need to install additional plants. By operating at a high air temperature, the spray rate can be increased to maximum and the coating time reduced to a minimum. Under these circumstances, the amount of heat used and possibly the amount of coating will be more than the minimum required, but the additional cost per batch is likely to be acceptable in comparison to the cost of an additional installation.

These cases are extreme examples but show that thermal efficiency may not be a major factor in the efficiency of the total process. The importance of thermal efficiency will vary with the type, cost, and availability of the fuel used. It should therefore always be considered, but only in relation to other factors. The most important factor is that the drying capacity be matched to the spray rate and not the reverse.

VII. CONCLUSIONS

In selecting equipment for applying a sustained release coating, or wherever even distribution of coating is an important factor, there are three basic requirements. The system must give efficient mixing and this must be achieved without damage to the tablets. The spray system must give good distribution of the solution with a regular droplet size. The drying system must enable both of these factors to be exploited to the maximum.

REFERENCE

1. **Leaver, T. M., Shannon, H. D., and Rowe, R. C.,** A photometric analysis of tablet movement in a side-vented perforated drum (Accela-cota), *J. Pharm. Pharmacol.,* 37, 17, 1985.

Chapter 5

MICROENCAPSULATION BY SOLVENT EVAPORATION AND ORGANIC PHASE SEPARATION PROCESSES

Jones W. Fong

TABLE OF CONTENTS

I. INTRODUCTION

Solvent evaporation and organic phase separation are two processes by which microencapsulation can be readily performed in the laboratory without the need of specialized equipment. The major difference between the two procedures is that solvent evaporation is an aqueous system whereas organic phase separation is nonaqueous. In this respect, the two processes are complementary in that the limitation of one method may be circumvented by using the other. Therefore, many core materials can be microencapsulated using one of the two processes. Selecting the appropriate method requires consideration of the physicochemical properties of the core material in conjunction with the characteristics of the process.

The following objectives should be considered in developing a microencapsulation procedure:

1.   High yields of microspheres, free of agglomeration
2.   High efficiency of encapsulation of the core material
3.   Reproducible release profiles from batch to batch
4.   Ability to modify in vitro release rates by varying process parameters in order to prepare microspheres with the desired in vivo release characteristics

In the first part of this chapter, the general procedures will be outlined and the advantages and limitations of each method discussed. This will then be followed by an in-depth examination of various aspects pertinent to each process.

II. GENERAL CONSIDERATIONS

A. Experimental Procedures

The general procedure for the solvent evaporation process is outlined below:

1.   The core material is either dispersed or dissolved in a solution of the wall-forming polymer. The polymer solvent can be any volatile, water-immiscible organic solvent.
2.   The organic phase is mixed with an aqueous solution containing a surfactant to form an oil-in-water (o/w) emulsion. The oil droplets of the emulsion contain the core material, wall-forming polymer, and the polymer solvent.

3. The polymer solvent is then removed from the dispersion of oil droplets to yield hardened microspheres consisting of the core material and the polymer. Solvent removal can be done at atmospheric or reduced pressure. The temperature of evaporation may also be varied. Solvent removal can also be accomplished by nonsolvent extraction, freeze-drying, or spray-drying.
4. The hardened microcapsules are collected by filtration or centrifugation.

The procedure for the organic phase separation process is as follows:

1. The core material is either dissolved or, preferably, dispersed in a solution of the wall-forming polymer.
2. The polymer is induced to phase-separate into a polymer-rich liquid phase which engulfs the core material to form embryonic microcapsules. Phase separation can be effected by (1) addition of a polymer nonsolvent, (2) addition of a second polymer incompatible with the wall-forming polymer, (3) lowering the temperature to decrease the solubility of the wall-forming polymer, or (4) solvent evaporation from a nonaqueous system. At this point, the microcapsules are usually too soft to collect.
3. A large excess of nonsolvent for the polymer is added to harden the microcapsules which can then be filtered.

B. Advantages and Limitations

Both of these processes share an advantage in that many polymers can be used, provided they are soluble in an organic solvent. Therefore, a large number of polymers with different physical and chemical properties can be evaluated as candidates for the capsule wall. Additionally, the polymer can be purified to the required degree prior to the fabrication step.

Another advantage is that both processes are suitable for (but not limited to) small-scale laboratory preparation of microspheres. This is especially important when only a limited amount of experimental or expensive polymer or drug is available. Microsphere batches of 0.5 to 1.0 g have been routinely prepared in laboratories, as well as scale-ups to 10 to 100 g.

Each of these processes has its limitations. Since the solvent evaporation method involves an aqueous emulsion, the core material is limited to those which have low solubility in water. The organic phase separation procedure, being nonaqueous, does not have this restriction and can be used for both water-soluble and water-insoluble core materials.

However, the organic phase separation method does have its own limitation. Core material used in this method must be insoluble in the nonsolvent employed for hardening the wall of the embryo microcapsules. Since the microspheres in the solvent evaporation process are hardened by solvent removal and not by addition of polymer nonsolvent, core material used in this process does not share this limitation of the organic phase separation method.

As noted above, only water-insoluble core material can be encapsulated in the solvent evaporation process, whereas both water-soluble and water-insoluble core material can be used in the organic phase separation process. In this respect, the organic phase separation method may appear to be the more versatile of the two processes.

However, from an operational point of view, the solvent evaporation process is easier to conduct and is less prone to agglomeration of the microspheres. The organic phase separation procedure requires more critical adjustment of the process parameters and is more susceptible to agglomeration of the microcapsules. Another factor in favor of the solvent evaporation method is that it is easier and more economical to handle large volumes of water than the large volumes of organic nonsolvent often required in the organic phase separation process. Considering all of the above, the solvent evaporation method would be the preferred process, especially by one who has limited experience in microencapsulation.

In summary, the major advantage of the solvent evaporation process lies in its simplicity and ease of operation. However, organic phase separation is obviously the process of choice for microencapsulating water-soluble core materials.

C. Microsphere Structure

The terms "microcapsules" and "microspheres" have been used interchangeably in the literature. In this chapter, they will be used in a more restrictive manner. The term "microcapsules" will be used to indicate heterogeneity. These include reservoir microcapsules (central core surrounded by a discrete coating) and monolithic spheres containing dispersed core particles. The term "homogeneous microspheres" will refer to monolithic spheres containing dissolved core material. The term "microspheres" will be used in a more general sense to denote either "heterogeneous microcapsules" or "homogeneous microspheres". Whether heterogeneous microcapsules or homogeneous microspheres are formed depends on the process employed and the solubility of the core material in the polymeric matrix.

To simplify the discussion, it will be assumed that if the core material is insoluble in the polymer solution it will probably be insoluble in the wall-forming polymer. Solubility of the core material in the polymer solvent will be assumed to suggest that the core material is also soluble in the polymeric matrix. In reality, these assumptions may not always be correct and some intermediate of the two structures may actually be obtained.

Heterogeneous microcapsules will usually be formed in the organic phase separation process because a dispersion of core particles is microencapsulated.

In the solvent evaporation process, the product will be either heterogeneous microcapsules or homogeneous microspheres, depending on the solubility of the core material in the polymer solution. If the solvent is selected to dissolve the polymer but not the core material, the structure will be heterogeneous with discrete core particles distributed throughout the resultant microcapsules. On the other hand, if the solvent is selected to dissolve both the polymer and the core material, homogeneous microspheres will be formed.

However, if the core material crystallizes during solvent evaporation to form discrete drug-rich domains within the polymeric matrix, the microspheres will then have a heterogeneous structure. Experimentally, this was found to be the case with the microencapsulation of several drugs.[1-3] This phenomenon will be discussed in further detail in Section III.C.

Microsphere structure has been examined using differential thermal analysis, differential scanning calorimetry, X-ray diffraction analysis, and scanning electron microscopy (SEM).[1,2,4,5] Using these techniques, it has been shown that the actual structure is dependent on the drug loading and the particular polymer used.

For example, progesterone did not crystallize but was molecularly dispersed in poly(DL-lactide) microspheres.[2] Differential thermal analyses of these microspheres exhibited no transition at the 131°C melting point of progesterone. The glass transition temperature of the polymer was not altered by the drug, indicating no interaction between the polymer and the drug.

In the case of cellulose acetate butyrate microspheres, the weak thermal event at 164°C for the polymer disappeared when progesterone was encapsulated in it.[1] This indicates that a solid solution was formed between the polymer and the drug.

With polystyrene microspheres, a solid solution was also formed at progesterone levels up to 35%. Above this drug loading (50 to 65%), a small amount of the progesterone crystallized in the polystyrene matrix, as suggested by the thermal event at 126°C for progesterone.[5]

Mathematical treatments on how drug release is affected by microsphere structure and other factors have been reviewed.[6-10] For monolithic microspheres containing drug either dissolved or dispersed in the polymeric matrix, the amount of drug release initially is proportional to the square root of time. For heterogeneous microcapsules containing dispersed drug particles, this drug release is continued with the $t^{1/2}$ dependence during most of the

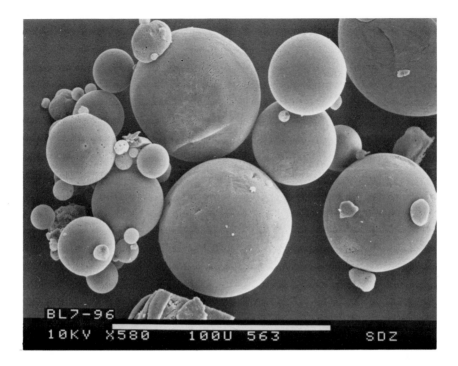

FIGURE 1.   SEM of hydrocortisone acetate microcapsules. (Courtesy of Visscher, G., Sandoz Research Institute.)

release duration. However, this is true only for the first 40% of drug release from homogeneous microspheres containing dissolved drug. Between 40 and 60% release there is a progressive change from $t^{1/2}$ to exponential dependence. The final 40% of drug is released in an exponential manner.

The product isolated from the solvent evaporation or the organic phase separation process will generally be monolithic microspheres containing dissolved or dispersed drug particles. It may be possible to prepare reservoir microcapsules having a central core coated with the polymer by using the two processes in sequence, as suggested in Section III.C. In this case, the rate of drug release is constant and is independent of time (zero-order release).

A SEM of typical microcapsules prepared by the solvent evaporation process is shown in Figure 1. In Figure 2 the microcapsules were fractured to expose the internal structure.

Recently, SEM was used to monitor the changes in the internal morphology of biodegradable microcapsules which were injected into rats.[11,12] The changes observed with SEM indicate that erosion of the internal matrix of the microcapsules occurred first. External erosion of the surface of the microcapsules was not significant until after internal erosion was well advanced.

D. Selection of Microencapsulation Process

In selecting between the two microencapsulation processes, the primary criterion is the solubility of the core material in water.

If the core material is water-insoluble, either the solvent evaporation or organic phase separation method can be considered. In general, the solvent evaporation procedure is preferred because of its ease of operation and lesser tendency toward agglomeration of the microspheres.

For a water-soluble core material, only the organic phase separation procedure is suitable. There are a number of methods for inducing phase separation of the wall-forming polymer and these will be examined in the latter part of this chapter.

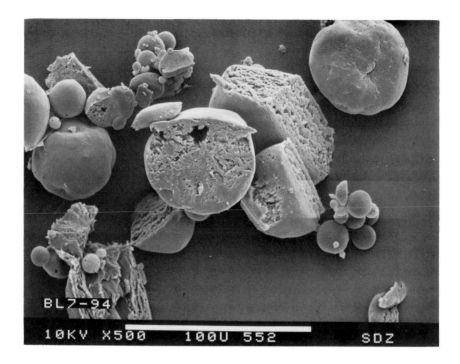

FIGURE 2.   SEM of hydrocortisone acetate microcapsules fractured to expose internal matrix. (Courtesy of Visscher, G., Sandoz Research Institute.)

III. SOLVENT EVAPORATION PROCESS

A comparison of the general procedures, advantages, and limitations of the two microencapsulation processes was presented in the preceding section. This portion of the chapter will deal with various aspects which are pertinent to the solvent evaporation process. Some of these aspects, e.g., the wall-forming polymer, are also applicable to the organic phase separation method.

A. Emulsifier

The emulsifiers commonly employed in these processes are macromolecular, hydrophilic colloids such as gelatin and polyvinyl alcohol. Anionic and mixed hydrophilic colloid-nonionic surfactants have also been used.

Selecting an effective emulsifier is important because its primary role is to prevent agglomeration of the microspheres during the fabrication process. Agglomeration is frequently the initial difficulty encountered in developing any procedure for microencapsulation.

In these emulsion processes, the surfactant must first stabilize the o/w emulsion against coalescing into larger droplets with subsequent "breaking" of the emulsion. As the solvent is being evaporated, the emulsifier must continue to maintain these oil droplets in their spherical configuration and stabilize them against agglomeration until solvent removal is completed. In this way the hardened microspheres can be isolated as discrete particles. Since the emulsifier plays such a critical role, the selection of an effective emulsifier is important.

The hydrophile-lipophile balance (HLB) method is only useful as a general guide for the selection of surface-active agents, based on the type of emulsion desired.[13] For example, materials with high HLB values (8 to 18) are o/w emulsifiers and those with low HLB values (3 to 6) are the water-in-oil (w/o) type. Since these processes are based on o/w emulsions, high HLB emulsifiers should be used. However, the HLB method does not indicate the efficiency (required concentration) nor the effectiveness (stability of the emulsion) of the surfactant. These aspects have to be determined experimentally. Therefore selecting an appropriate emulsifier requires more than just a consideration of HLB values.

A number of nonionic and anionic emulsifying agents with high HLB values (for o/w emulsions) were evaluated for their effectiveness in the solvent evaporation microencapsulation process.[14] These included nonionic emulsifiers such as BRIJ®, Chremophor®, MYRJ®, and Tween®. None of these surfactants provided a sufficiently stable emulsion. Alkali salts of carboxylic acids were also considered. Of those examined, the fatty acid salt emulsifiers, preferably sodium or potassium oleate, were found to be suitable for the preparation of microspheres from biodegradable polymers.

The use of sodium oleate as the emulsifier gave high yields (74 to 96%) of biodegradable microspheres, essentially free of agglomeration.[3,14-16] High efficiency of drug encapsulation (80 to 99%) was routinely found. Core loading as high as 50% could be obtained along with reproducible controlled release. Most of the microspheres were less than 150 μm in diameter. This size consideration is important when the microspheres made with biodegradable polymers are intended for injectable pharmaceutical applications. The 150-μm diameter represents the maximum size which would pass through a 20-gauge needle.

Since fatty acids are endogenous lipids, residual amounts in the microspheres should be pharmaceutically acceptable. Microspheres prepared by this process using sodium oleate as the emulsifier caused no abnormal tissue response when injected intramuscularly into rats and dogs.[3]

The use of fatty acid salt emulsifiers in a solvent evaporation microencapsulation process satisfies the following prerequisites for parenteral application:

1. Discrete microspheres, free of agglomerates, with maximum diameter of 150 μm
2. Reproducible release rate which is significantly slower than the nonencapsulated drug
3. All ingredients in the microspheres (including residual amounts) nontoxic and pharmaceutically acceptable

Microencapsulation of medicaments by a solvent evaporation process was initially reported by Morishita et al.[17] Hydrophilic colloids or surface-active agents with an HLB of at least 10 were used to form the o/w emulsions of their process. Aqueous solutions containing 0.5 to 2.0% w/v of gelatin, polyvinyl alcohol, or hydroxyalkyl cellulose are examples of hydrophilic colloids used as the emulsifiers. Examples of surface-active agents are aqueous solutions containing 0.1 to 1.0% w/v of anionic or nonionic surfactants with an HLB of at least 10. Mixed solutions of both hydrophilic colloid and surface-active agent were also used.

The effect of using different types of gelatin emulsifier on drug incorporation was reported by Wakiyama et al.[18] For basic amine drugs such as dibucaine, the efficiency of drug incorporation (drug content assayed ÷ theoretical drug content) was affected by the type of gelatin used as the emulsifier. Although the yields were about the same (56%), the efficiency of drug incorporation was only 31% using acid-processed gelatin and was increased to 81% with alkaline-processed gelatin as the emulsifier. The difference in results was related to the pK_a of the drug and the pH of the gelatin solution. This aspect will be discussed in greater detail in the next section.

Low yields of small microspheres (diameter less than 150 μm) were sometimes obtained with these emulsifier systems. As mentioned above, this size consideration is important when the microspheres made with biodegradable polymers are intended for injectable pharmaceutical applications. In some cases, the low yield of small microspheres was attributed to coarse powder resulting from aggregation of smaller microspheres. Microspheres with diameters of 177 to 595 μm were reported by Jaffe[19] using an anionic surfactant, sodium lauryl sulfate. The diameters of many microspheres exceeded 150 μm in some examples[17] which used gelatin as the emulsifier. Similar results using gelatin as the emulsifier were also observed by Wakiyama et al.[20] Yields of microspheres with diameters less than 149 μm varied from 18 to 73%. Using a mixed hydrophilic colloid-nonionic surfactant, Yolles et al.[21] obtained microspheres of 250 to 420 μm in diameter with some larger agglomerates.

Polyvinyl alcohol has been used as the emulsifier in some investigations. Beck et al.[22] obtained microspheres with diameters of 10 to 250 μm. Using the same emulsifier, Bissery et al.[4] isolated microspheres with a range in diameter of 15 to 600 μm, the size being a function of the agitation rate.

As previously mentioned, most of the microspheres prepared by the process using sodium oleate as the emulsifier were less than 150 μm in diameter.[3,14-16]

B. Aqueous Phase

The active ingredient in the aqueous phase is the emulsifier. Due to its important role in these microencapsulation processes it was discussed separately in the preceding section. Other aspects of the aqueous phase will be discussed in this section.

The pH of the aqueous phase may have to be adjusted for optimum performance of the emulsifier. For example, an alkaline pH is required for anionic surfactants, such as the fatty acid salts.[14] The pH of the aqueous phase may also need to be adjusted for the core material. This is due to the effect of pH on the solubility of the core material in the aqueous vehicle. For example, it is more desirable to microencapsulate a drug in its nonionized free base form. This would minimize loss due to solubility of the ionized drug in the aqueous phase of the emulsion. The required alkaline pH can be satisfied by the addition of alkali or an alkaline buffer solution.[14]

In the above section it was noted that higher efficiency of drug incorporation in the microencapsulation of a basic amine drug was obtained using alkaline-processed gelatin instead of acid-processed gelatin.[18] This can be explained when the pH of the gelatin solution, the pK_a of dibucaine, and the partition coefficient of the drug between the aqueous and organic phases at a given pH are considered. As the pH of the aqueous medium is increased from 3.8 in acid-processed gelatin to 7.5 in alkaline-processed gelatin, dibucaine (pK_a = 1.6 and 8.3) changes from a completely ionized form (more water-soluble) to a less ionized form (less water-soluble). This is reflected in the partition coefficient which is decreased by an order of magnitude due to the increase in pH. When dibasic sodium phosphate is added to the aqueous gelatin solution, the pH is further increased to 8.6, which is slightly above the second pK_a of dibucaine. The drug is now in the predominantly nonionized free base form. This results in a further decrease in the partition coefficient with a corresponding increase in efficiency of drug incorporation, as evidenced by the higher drug content obtained.

The pH of the aqueous phase may also have a significant effect on the release profile of the core material. The addition of sodium hydroxide to the aqueous phase of the emulsion prior to the solvent evaporation step enhanced the release of some amine drugs from polylactide microspheres.[15,16] The rate of drug release was found to be dependent on the amount of base added: it was faster at the higher level of sodium hydroxide and slower at the lower level.

The aqueous phase may also contain an inorganic salt to lower the solubility of the core material by the salting-out effect.

Another method of minimizing the partitioning of the core material into the aqueous phase of the emulsion is to saturate the aqueous phase with the core material. This would help maintain the desired concentration of the core material in the oil phase of the emulsion.[23,24] For example, when the tetracaine level in the aqueous phase was increased from 0 to 0.75%, the encapsulation efficiency for this drug was increased from 45 to 99%.[23] To conserve the drug, the aqueous phase saturated with the core material can be recycled to prepare another batch of microspheres.

C. Core Material

A wide variety of core materials has been microencapsulated by the solvent evaporation process. These include contraceptive steroids,[1,21,22,25] local anesthetics,[18,20,23] antibiotics,[26] antineoplastics,[27] insulin,[28] other pharmaceuticals,[3,14,17] activated carbon,[29,30] pesticides,[19,31,32] reagents,[33] and inks.

The major prerequisite for any core material to be used in the solvent evaporation process is that its solubility in the aqueous phase be sufficiently low. Core materials with some water solubility tend to partition from the organic phase into the aqueous phase prior to completion of solvent evaporation, resulting in low or no drug loading.

In the previous section it was stated that a drug may be microencapsulated in its nonionized free base form. The purpose of doing this was to minimize loss due to solubility of the ionized drug in the aqueous phase of the emulsion. Acid salts of amine drugs with appreciable solubility in water were converted to their less soluble, free base form by the addition of the stoichiometric amounts of alkali. These basic drugs with aqueous solubility of about 0.02 mg/mℓ or less were then microencapsulated to give reproducible, controlled release.[3] Similarly, hydrocortisone acetate with aqueous solubility of less than 0.02 mg/mℓ was also microencapsulated to give a release profile with significantly retarded release.[3] However, it has been reported that incorporation of a basic medicament, e.g., thioridazine free base, can enhance the degradation of the wall-forming polymer.[34]

However, when steroids with aqueous solubility of 0.08 mg/mℓ or higher (e.g., cortisone, cortisone acetate, prednisone, and prednisolone) were microencapsulated, considerable amounts of free drug crystals were also present.[3] Droplets of the emulsions (prior to the solvent evaporation step) containing these more soluble steroids were examined with a polarized light microscope. During evaporation of the organic solvent from these droplets, iridescent crystal growth was observed on the surface of the microspheres and in the aqueous phase. This is illustrated in a SEM in Figure 3. The resultant microspheres with partially encapsulated drug particles may account for the high initial release observed with these more soluble steroids. The release patterns of these microspheres were equal to or only slightly slower than the nonencapsulated drugs.

The formation of free drug crystals during the fabrication process was also observed by Bissery et al.[1] and Thies and Bissery.[2] They reported that if an emulsifier, such as polyvinyl alcohol, is left in the system until the microspheres are completely formed, some free progesterone drug crystals will form in the aqueous phase. It was suggested that the emulsifier assists the nucleation and growth of free crystals. In order to prevent or reduce the amount of free drug crystals, the aqueous emulsifier solution was removed from the system prior to complete evaporation of the polymer solvent and was replaced with surfactant-free water for the remaining solvent evaporation step. However, there is a critical time period for doing this. If the emulsifier is removed too soon, the embryo microspheres will aggregate. If this is done too late, free crystals will form. This critical time period is dependent on factors such as polymer type, polymer molecular weight, initial polymer concentration in the organic phase, agitation rate, and solvent evaporation rate.

A slightly different approach was used by another group of investigators to avoid the formation of free crystals in the microencapsulation of progesterone.[22,35] When half of the polymer solvent was evaporated from the emulsion, the suspension was centrifuged. The supernatant was decanted and the microspheres were suspended in deionized water. The microspheres were then filtered through a fritted glass filter under vacuum and continuously washed with water until the remaining solvent was completely removed. It was not indicated whether there is a critical time period for the extraction step, analogous to the method of Bissery et al.[1] and Thies and Bissery[2] in which the second half of the solvent evaporation process was conducted in surfactant-free water. Other methods to reduce partitioning of core material into the aqueous phase of the emulsion were elaborated on in the preceding section.

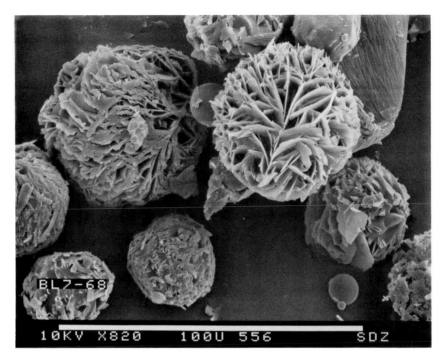

FIGURE 3. SEM showing crystal growth of cortisone acetate on surface of microcapsules. (Courtesy of Visscher, G., Sandoz Research Institute.)

Although much of the discussion presented thus far in this section has been concerned with circumventing the aqueous solubility of the core material, one of the earliest microencapsulation procedures by solvent evaporation was developed for encapsulating an aqueous solution as the core material.[36] This involved a more complex emulsion system (water/oil/water) rather than the simple o/w emulsion described in Section II.A. Appropriately, this has also been called the complex emulsion method in the literature.[37] However, the basic principle is still that of solvent evaporation from an emulsion system.

A companion patent[38] to the one cited above indicated that removal of solvent could also be accomplished by nonsolvent extraction. This involves extracting the polymer solvent (cyclohexanone) with a liquid (methanol/water) which is miscible with both water and the polymer solvent but is a nonsolvent for the polymer (vinylidene chloride/acrylonitrile copolymer).

However, the use of the complex emulsion process does not always circumvent the problem of core material partitioning into the aqueous phase of the emulsion. This was indicated in another related patent.[24] A water-retaining substance, such as glycerol or ethylene glycol, was included in the aqueous core to reduce migration of the aqueous core solution through the capsule wall during the solvent evaporation step.

A particularly novel application of microencapsulation by solvent evaporation was exemplified by microspheres containing reagents capable of detecting toxic compounds.[33] Instead of the usual application in which core material is released from the polymeric matrix, these microspheres were designed to retain the encapsulated reagent within the polymeric wall. When exposed to insecticides or pesticides, these microspheres produce a color change caused by the organophosphorus compounds penetrating into the microspheres and reacting with the encapsulated azo reagent.

Another interesting application was the use of air-filled microcapsules as an opacifier in a latex paint system. These were prepared by microencapsulating a volatile polymer nonsolvent with subsequent removal of the nonsolvent by spray-drying to form the hollow microspheres.[39]

It is also possible to prepare microspheres of core material (i.e., microspheres of drug only without the polymeric capsule wall) or placebo microspheres of polymer alone (containing no core material).[14] The microspheres of drug (without polymer) may be used as spherical core material in a subsequent microencapsulation step. The resultant product would be reservoir microcapsules having a central core coated with a polymeric wall. These would be different from the usual monolithic microspheres in which the core material is either dissolved or dispersed throughout the polymeric matrix.

The release profiles from the two types of microspheres are different. The rate of release of core material from reservoir microcapsules is constant and is independent of time (zero-order release). In the case of monolithic microspheres, the rate of release decreases with time. A brief discussion on the effect of microsphere internal structure on drug release was provided in Section II.C.

As mentioned at the beginning of this section, the major prerequisite for the core material is that it must have low aqueous solubility. The solubility of the core material in the organic solvent for the wall-forming polymer is less important, although the internal structure of the microspheres is affected. The microspheres will have a homogeneous structure if the core material is soluble in the polymer solution. If the core material is insoluble in the organic solvent, the product will be heterogeneous microcapsules.

For preparing heterogeneous microcapsules, the particle size of the core material should be reduced by micronization of the core material prior to dispersing in the oil phase or by mixing a dispersion of the core material and the organic solvent with an ultrasonic homogenizer prior to the emulsification step. It may be preferable to homogenize the core material dispersion in the absence of the wall-forming polymer because of the possibility that the molecular weight of some polymers can be reduced by the high shearing force of the homogenizer.

As core loading is increased, there is a corresponding decrease in the relative amount of polymer in the wall of the microspheres which can act as a barrier to the release of the core material. Therefore, an increase in core loading can be expected to cause an increase in the rate of release of the core material. Release patterns of microspheres containing dibucaine,[18] butamben, and tetracaine[23] all exhibited a higher release of drug when core loading was increased. A similar effect was observed with microspheres loaded with ketotifen and hydrocortisone acetate.[3]

The in vitro release of progesterone from microspheres containing 75% of the steroid was much higher than those with 50% drug loading.[40] This was also reflected in the progesterone serum level when these two formulations were injected into baboons.

There is often a practical upper limit to the core loading. This limit is dependent on the core material. For example, when chlorpyrifos was microencapsulated above the 40% level, the pesticide plasticized the wall-forming polymer and caused severe agglomeration of the microspheres.[32] However, a number of drugs, e.g., thioridazine and ketotifen, were successfully microencapsulated at the 50 to 60% levels.[3,14-16]

D. Wall-Forming Polymer

The polymer used for forming the wall of the microspheres must be insoluble in the aqueous continuous phase. In addition, it must also be soluble in a water-immiscible solvent. The latter requirement is easily satisfied since many water-insoluble polymers are soluble in methylene chloride. This polymer solvent is frequently used in these solvent evaporation processes because of its relative ease of evaporation.

The particular polymer may be selected on the basis of various considerations such as physical properties, biodegradation rate, and the effect of the polymer on release profiles of the core material.

A number of synthetic polymers have been used in these emulsion processes. Ethyl cellulose was the wall-forming polymer used for microencapsulating activated charcoal.[29,30] Both polystyrene and cellulose acetate butyrate were used to prepare microspheres of progesterone.[1]

Polymers with inherent tackiness caused severe agglomeration of the microspheres after isolation and drying.[32] Examples of these are polyurethane, poly(ethylene-vinyl acetate), poly(styrene-butylene-ethylene-styrene), poly(styrene-isoprene-styrene), and chlorinated polyethylene.

Polymethyl methacrylate was selected as the polymer for microencapsulating azo reagents which produce color changes when exposed to insecticides and pesticides.[33] This polymer was chosen because its solubility parameter is similar to those of the organophosphorous compounds to be tested. Compounds of dissimilar solubility should not penetrate these reagent microspheres and therefore would not cause any color change.

In recent years, considerable attention has been focused on biodegradable polymers for preparing microspheres of pharmaceutical agents. Using these biodegradable microspheres in parenteral applications obviates the need for surgical removal of the microspheres after delivery of the drug. Examples of biodegradable polymers which have been used in these solvent evaporation processes include homopolymers and copolymers of lactic acid,[1-5,14-16,21-23] glycolic acid,[3,15,16,25,41] β-hydroxybutyric acid,[1,4,5] ε-caprolactone,[1,3,15,16] and alkylene glycol esters of the Krebs cycle acids.[14]

1. Polymer Composition

The composition of the wall-forming polymer can affect the release patterns of the core material from the microspheres. This may be due to a number of polymer properties such as biodegradation rate, monomer ratio, crystallinity, glass transition temperature, permeability, and solubility of the core material in the polymer.[42-44]

For polymers with slow degradation rates, the predominant mode of release of the core material is diffusion through the polymeric matrix. With faster-degrading polymers, release from microspheres occurs initially by diffusion, with subsequent contribution resulting from concomitant release of core material due to erosion of the degrading polymer. Therefore, one would expect that release of core material would be faster from microspheres prepared with fast-degrading polymers than those prepared from polymers with slower degradation rates. Since copolymers of lactide-glycolide[25,41] and lactide-ε-caprolactone[45] have faster degradation rates than poly(DL-lactide), microspheres made with these copolymers should release core material faster than those from poly(DL-lactide).

a. Homopolymers

When polylactides were used to prepare medicated microspheres, the release profiles were different from what would be theoretically expected. The release of both thioridazine and ketotifen from microspheres made with poly(L-lactide) was much faster than those prepared with poly(DL-lactide).[3,15,16] These results are not consistent with the previous discussion, since the degradation of poly(L-lactide) is slower than that of poly(DL-lactide). These results are also not expected on the basis of polymer crystallinity. Diffusion of drug through polymeric membranes was reported to be reduced by an increase in polymer crystallinity.[43] Therefore, the more crystalline poly(L-lactide) is expected to be slower releasing than the amorphous poly(DL-lactide). Apparently there are factors other than polymer degradation and crystallinity which can impact the release patterns of core material. Channels or pores can also have a significant effect on drug release, as well as other polymer properties mentioned in the preceding section.

Poly(β-hydroxybutyrate) is another homopolymer which has been evaluated for fabrication of biodegradable microspheres. Core materials encapsulated with this polymer include progesterone, cyclohexylchloroethyl nitrosourea (CCNU), 5-fluorouracil, red Sudan dye, and radiopaque agent.[1,4,5] The surfaces of CCNU-loaded microspheres prepared from this polymer were highly porous when examined by SEM. The interior of these microspheres also contained many pores and voids, according to the transmission electron micrographs. In contrast, the surfaces of microspheres prepared from poly(DL-lactide) were smooth and almost free of defects. The physical differences between these two samples of microspheres were reflected in their in vitro release patterns. CCNU-loaded microspheres prepared with poly(β-hydroxybutyrate) released essentially all the drug content within 1 day, whereas the duration of in vitro release was 7 days for those made with poly(DL-lactide).

b. Copolymers

As stated in Section III.D.1, copolymers of lactide-glycolide and lactide-ε-caprolactone have faster degradation rates than polylactide and therefore should release the core content faster. As expected, the drug release of thioridazine-loaded microspheres prepared from both 75:25 poly(DL-lactide-co-glycolide) and 75:25 poly(L-lactide-co-glycolide) was faster than that of those made with poly(DL-lactide).[3] However, the drug release from both microsphere samples was slower than that of those made with poly(L-lactide), even though the biodegradation rates of the two copolymers were faster than that of poly(L-lactide). These anomalous results with poly(L-lactide) were mentioned previously.

Similar release patterns were observed with ε-caprolactone copolymers. The release of thioridazine from microspheres prepared from both 90:10 poly(DL-lactide-co-ε-caprolactone) and 80:20 poly(L-lactide-co-ε-caprolactone) was faster than that of those from poly(DL-lactide).[3] As with the above lactide-glycolide copolymers, thioridazine was also released slower from these ε-caprolactone copolymers than from microspheres made with poly(L-lactide), even though the latter degraded slower than the two caprolactone copolymers.

Microcapsules containing norethisterone were fabricated with a series of radiolabeled DL-lactide-glycolide copolymers.[25,46] These copolymers had approximately the same molecular weight but were synthesized with different lactide:glycolide mole ratios. The biodegradation of the copolymers in rats was followed by measuring the loss of radioactivity. The rate of biodegradation was higher when the mole percentage of glycolide was increased from 4 to 26%.

Norethisterone-loaded microcapsules prepared with several DL-lactide-glycolide copolymers were injected intramuscularly into baboons and the serum level of the drug was assayed. The duration of norethisterone release in baboons corresponded closely to the polymer residence time based on the biodegradation studies in rats. Furthermore, the serum norethisterone level revealed a biphasic release of the steroid. This indicated an initial release due to diffusion followed by a later release which coincided with the biodegradation of the copolymer. This later release occurred earlier with the higher glycolide content copolymer. These data suggest that it possible to modify release profiles by selecting wall-forming materials with different biodegradation rates.

2. Polymer Molecular Weight

An increase in the molecular weight of the wall-forming polymer leads to a decrease in the release pattern of the core material. This was demonstrated with microspheres containing butamben and tetracaine which were microencapsulated with poly(DL-lactide).[20] Polymers with molecular weights of 9100, 17,000, and 25,000 were used in this study.

The extent of the molecular weight effect diminishes at the higher molecular weight levels of the polymer. For example, thioridazine-loaded microspheres, prepared from poly(DL-lactide) with molecular weights of 51,000 and 170,000, showed the same release patterns during the first 70% of drug release. However, the remaining drug content was released somewhat slower from the polymer with the higher molecular weight of 170,000.[3]

Another example of the polymer molecular weight effect was exhibited by norethisterone-loaded microcapsules which were sterilized by gamma irradiation. The inherent viscosity of poly(DL-lactide-co-glycolide) decreased from about 0.8 to 0.5 dℓ/g after treatment, indicating a decrease in the molecular weight of the wall-forming copolymer.[46] This was reflected by an increase in the biodegradation rate of the copolymer and led to the expected faster release of the core material (see Section III.D.1).

3. Initial Polymer Solution Concentration

The initial concentration of the wall-forming polymer solution (polymer-solvent ratio) in the organic phase of the emulsion can have a significant effect on the in vitro release patterns of the core material. Release of core material from the microspheres was found to decrease when the initial concentration of the polymer solution was increased prior to the emulsification step.[3] Experimentally, this was done by dissolving the same amount of the wall-forming polymer in decreasing volumes of the organic solvent. The extent of this effect may be dependent on the drug.

In the preparation of 43% thioridazine-loaded microspheres, increasing the initial concentration of the polymer solution from 5 to 10% caused a drastic decrease in the in vitro release of the drug.[3]

The effect of initial polymer solution concentration on drug release was much less with hydrocortisone acetate microcapsules prepared at both 23 and 40% drug loadings.[3] As expected from the previous discussion on core loading (Section III.C), the 23%-loaded microcapsules released hydrocortisone acetate much slower than those prepared at the higher 40% drug level. At each drug level there was only a slight decrease in drug release with increases in the initial concentration of the polymer solution.

The difference in the extent of the effect of initial polymer solution concentration on drug release between thioridazine and hydrocortisone acetate microspheres may be due to more than just the fact that they are different drugs. It may also be due to the difference in the solubility of the two drugs. Thioridazine, when completely dissolved in the polymer solvent, formed homogeneous microspheres. Heterogeneous microcapsules were formed with hydrocortisone acetate because the drug was insoluble in the polymer solution. It may be that the effect of initial polymer solution concentration on drug release is greater for core material which is soluble in the polymer solution than for one which is insoluble.

E. Polymer Solvent

The organic solvent for the wall-forming polymer must be immiscible or only partly soluble in water. Its boiling point must be lower than that of water if solvent removal is to be accomplished by evaporation. The boiling point of the solvent is of less importance if nonsolvent extraction is used to remove the polymer solvent.

The solvent commonly used in these processes is methylene chloride because of its ease of removal and its excellent ability to dissolve many polymers. Other solvents which have been used include chloroform, carbon tetrachloride, ethylene chloride, ethyl ether, benzene, methyl acetate, and ethyl acetate. Mixtures of these solvents have also been used.

The relationship of the polymer solvent to the core material is important in that the internal structure of the microspheres may be affected. The microspheres will have a homogeneous structure if the polymer solvent also dissolves the core material. If the core material is insoluble in the polymer solvent, heterogeneous microcapsules will be formed.

F. Solvent Evaporation and Stirring Rate

Solvent evaporation can be accomplished by simply stirring the emulsion in an open beaker (a stream of nitrogen may be used to facilitate evaporation) or in a closed system under reduced pressure. Heat or reduced temperature can also be applied. Heat and vacuum, if used, should be moderate to minimize foaming of the emulsion during the initial part of the evaporation step. In Section III.C, two examples were provided of the use of interrupted solvent evaporation to minimize the formation of free crystals in the microencapsulation of progesterone.[1,2,22,35]

The key factor for obtaining batch-to-batch reproducibility of drug release is the use of identical encapsulation conditions. Ensuring identical conditions during the solvent evaporation step involves careful attention to controlling the amount of emulsifier, stirring speed, atmospheric or reduced pressure, solvent evaporation time, and temperature.[3]

The initial stirring rate should be adjusted such that oil droplets of the desired diameter are formed prior to the solvent evaporation step. As expected, smaller microspheres are formed by increasing stirring speed. During the evaporation step, stirring should be continued at a rate sufficient to prevent agglomeration of the microspheres until solvent removal is completed so that the hardened microspheres can be isolated as discrete particles.

The stirring speed may also have an effect on the quality of the microspheres. Stirring speeds of 800 to 1600 rpm were usually employed in most of the examples in which gelatin or polyvinyl alcohol was used as the emulsifier. However, a large number of small holes (1 to 10 μm) were observed on the surface of microcapsules prepared by rapid stirring (900 rpm).[47] No holes were detected in the SEMs of microcapsules prepared by slower stirring (500 rpm). These holes were considered to have a significant role in the release of encapsulated core materials. However, a comparison of the release patterns between the two types of microcapsules was not provided.

IV. ORGANIC PHASE SEPARATION PROCESSES

A. General Aspects

Microencapsulation by organic phase separation is dependent on a phenomenon called coacervation. This is a process in which a polymer solution is induced to separate into two phases, a dilute liquid phase and a condensed liquid phase. The dilute liquid phase is polymer-poor, whereas the condensed liquid phase is polymer-rich. The formation of the polymer-rich liquid phase, also called coacervate, is the basis for the different organic phase separation processes used in microencapsulation.

Organic phase separation of the wall-forming polymer can be induced by one of the following methods:

1. Addition of nonsolvent for the polymer
2. Addition of a second polymer incompatible with the wall-forming polymer
3. Change in temperature
4. Solvent evaporation (different from the aqueous emulsion process described in the first part of this chapter)

These methods all share a common feature in that the wall-forming polymer is desolvated to form a polymer-rich liquid phase during phase separation. As more of the polymer is gradually desolvated, the volume of the condensed polymer phase is progressively increased to a maximum point. During phase separation, the core material is occluded by the polymer-rich liquid phase to form the microcapsules.

1. Basic Requirements

Success in microencapsulation with the organic phase separation processes is dependent on at least three factors:

1. The condensed polymer phase must have an affinity for the core material. Otherwise

the product will be a mixture of nonencapsulated core material and polymeric particles containing little or none of the intended core material.

2.   It is important that the condensed polymer phase be sufficiently fluid to provide complete encapsulation. The phase separation should also be gradual to allow sufficient time for the polymer-rich liquid phase to deposit and spread evenly on the surface of the core material. This usually requires a slow addition of nonsolvent or a gradual temperature change to a dilute polymer solution. Higher polymer concentrations would yield a more rapid phase separation and also produce a condensed polymer phase which may be too viscous for satisfactory coating of the core material.

3.   The tendency of microcapsules to agglomerate during fabrication must be overcome. This is a problem frequently encountered in any microencapsulation study. The major cause for this problem is the presence of the solvent used to dissolve the wall-forming polymer. Until the solvent is completely removed, the microcapsules may not be sufficiently hardened and can adhere to each other. This is especially true for those organic phase separation processes in which the entire amount of polymer solvent is constantly present until the microcapsules are physically collected by filtration. The constant presence of the polymer solvent, even when diluted with polymer nonsolvent, prevents the microcapsules from being completely hardened and thus increases the possibility of the microcapsules coalescing together.

In comparison to the organic phase separation process, there is less tendency for agglomeration in the solvent evaporation process previously described. This is because the embryonic microspheres are initially in the form of the oil droplets of the emulsion. These oil droplets are physically separated from each other by the aqueous continuous phase and are stabilized against coalescence by the surfactant in the aqueous solution.

Agglomeration of microcapsules prepared by the organic phase separation processes can be minimized by a few techniques reported in the literature. These techniques usually involve some counteraction to the softening effect of the microcapsules by the polymer solvent. The various procedures will be individually considered in other sections of this chapter when the specific methods of phase separation are examined.

2. Wall-Forming Polymer

The general prerequisite for the wall-forming polymer in the organic phase separation processes is that it be soluble in an organic solvent. It must also have high affinity for the core material in order to be effective for encapsulation. Other requirements will be indicated when the specific method of phase separation is discussed.

Much of the discussion on wall-forming polymer, covered in Section III.D, is also applicable to these organic phase separation processes.

3. Polymer Solvent

For the organic phase separation processes, the requirements of the polymer solvent differ from those for the solvent evaporation process. The solvent does not have to be immiscible with water and its boiling point can be higher than that of water.

In the organic phase separation processes, it is preferable that the core material be dispersed rather than dissolved in the polymer solution. Therefore, the solvent should dissolve the polymer but not the core material. Actually, this requirement is necessary only under the experimental conditions existing just prior to phase separation of the wall-forming polymer. It is possible to encapsulate soluble core material provided it is precipitated out of solution before the wall-forming polymer has phase separated. This point is further elaborated below.

4. Core Material

Both water-soluble and water-insoluble core materials are candidates for microencapsulation by the organic phase separation processes. In fact, some of the earlier uses of the organic phase separation processes were for microencapsulating aqueous droplets.

The major requirement is that the core material be insoluble in the solvent used for preparing the wall-forming polymer solution. The reason for this is that the core material should be available for encapsulation prior to phase separation of the polymer. Otherwise, the product will be a mixture of nonencapsulated core particles and microcapsules containing little or no core material.

In some cases, it is possible to microencapsulate core material which is soluble in the polymer solvent, for example, when the selected polymer solvent is a better solvent for the polymer than for the core material. Since the core material in this case would have lower solubility in the solvent than the polymer, core particles would precipitate out of solution before the polymer begins phase separation. The core particles would then be present for encapsulation by the wall-forming polymer during the phase separation step.

Another circumstance in which a soluble core material can be microencapsulated is when phase separation of the wall material is initiated prior to introducing the core material to the system. This is only possible if the experimental conditions existing during phase separation minimize dissolution of the core material.

Specific examples in which soluble core material is microencapsulated will be provided when the individual methods of phase separation are discussed.

B. Phase Separation Induced by Nonsolvent Addition

In this method, the core material is dispersed in an organic continuous phase containing the dissolved wall-forming polymer. When a nonsolvent for the polymer is initially added to the dispersion, the solubility of the polymer is decreased, inducing the polymer to separate into a polymer-rich liquid phase. The latter is then deposited as a coating on the dispersed core particles. Further addition of the nonsolvent causes the coating to harden as a capsule wall encasing the core material. Thus the role of the nonsolvent in this method is to induce phase separation of the polymer and also to harden the wall of the newly formed microcapsules.

A typical example of this method, using ethyl cellulose as the wall-forming polymer, was described in a process developed by Reyes.[48] In this experiment, 5 parts of an aqueous solution containing dye components was dispersed in 1 part of a 5% solution of ethyl cellulose in toluene. The w/o dispersion contained 1% of sorbitan sesquioleate as the emulsifying agent. Petroleum ether (65 to 110°C) was slowly added to induce phase separation of the wall-forming polymer. Microcapsules were formed when the dispersed aqueous droplets were coated by the separated polymeric phase. After stirring and cooling to about 15°C, the supernatant was decanted. The microcapsules were hardened by a series of washings with petroleum ether containing decreasing amounts of toluene.

1. Wall-Forming Polymer

Typical polymers which have been used to form the capsule wall in the nonsolvent addition method include benzyl cellulose, ethyl cellulose, nitrocellulose, cellulose acetate, cellulose acetate butyrate, polystyrene, polyvinyl acetate, polyvinyl chloride, and polyvinyl stearate.[48,50]

The biodegradable polylactic acid has also been employed as the wall-forming polymer.[51,52] In a report by Itoh et al.,[51] microcapsules were prepared with a blend of ethyl cellulose and polylactic acid in the capsule wall. Sulfamethizole was suspended in a solution of polylactic acid and ethyl cellulose dissolved in ethyl acetate. Pentane was added dropwise to the suspension until the phase separation was completed. The microcapsules were washed with pentane, filtered, and dried.

Increasing the amount of polylactic acid in the capsule wall produced a corresponding increase in the release duration of sulfamethizole. However, at polylactic acid levels above 50%, the embryonic microcapsules tended to aggregate.

2. Control of Agglomeration

Agglomeration of microcapsules prepared entirely from polylactic acid was also observed by Fong.[52] This problem was circumvented by conducting the phase separation of the wall-forming polymer at $-65°C$. In this low-temperature process, a dispersion of drug particles in a solution of polylactic acid was cooled in a dry-ice bath. When the nonsolvent was added, the polymer phase separated as a viscous liquid and coated the drug particles to form microcapsules. The use of very low temperature caused the newly formed microcapsules to be sufficiently firm throughout the entire phase separation process; thus adhesion between the microcapsules was avoided.

Another method for minimizing coalescence of microcapsules was reported by Rowe.[50] The incorporation of a mineral silicate (e.g., talc) during the addition of the nonsolvent was found to minimize adhesion and coalescence of the microcapsules. The mineral silicate must be added intermittently during the phase separation to provide continuous protection against agglomeration. The talc particles on the surfaces of the microspheres provided a protective barrier against adhesion between the microcapsules. The addition of talc to avoid agglomeration of microcapsules was also used in some other patented processes.[53,54]

3. Polymer Solvent

The choice of solvent is dictated by the requirement that it should dissolve the intended wall-forming polymer but not the core material. The solvent must also be miscible with the nonsolvent for the capsule wall polymer.

El-Sayed et al.[55] showed that the choice of polymer solvent can have an effect on the release characteristics of the resultant microcapsules. Eudragit RLPM®, a methacrylate polymer, was used as the wall-forming material for the microencapsulation of riboflavin. Solutions of the polymer were prepared in both polar and nonpolar solvents. Phase separation was effected by the addition of petroleum ether as the nonsolvent for the wall material. When benzene was used as the nonpolar solvent for the polymer, phase separation resulted in the formation of large, well-defined polymeric droplets which coalesced as a thick uniform coating on the surface of the riboflavin crystals. When isopropanol was the polymer solvent, the coacervate droplets were tiny and the surface of the core particles was coated with a relatively thin film. This was reflected in the release profiles of riboflavin from tablets made with microcapsules using these two polymer solvents. The release rate was higher when isopropanol was used as the solvent.

4. Nonsolvent

The nonsolvent for the wall-forming polymer must meet several criteria. It must be a nonsolvent for the core material. In addition, the nonsolvent must be completely miscible with the polymer solvent. The nonsolvent should be relatively volatile or easily removed by washing with another volatile nonsolvent having all of the above requirements. The role of the nonsolvent is both to effect phase separation of the wall-forming polymer and to harden the newly formed microcapsules.

The choice of nonsolvent may have an effect on the fabrication of the microcapsules. Although both polar and nonpolar nonsolvents may be used in the low-temperature microencapsulation process,[52] polar nonsolvents (e.g., isopropanol, isobutanol) were preferred over nonpolar nonsolvents (e.g., heptane) for preparing the microcapsules. Furthermore, the combination of a polyhydric alcohol (e.g., propylene glycol) and isopropanol as the non-solvent mixture produced microcapsules with larger diameters (100 to 125 μm) than those derived from isopropanol alone (25 to 50 μm).

5. Core Material

Representative core materials which have been microencapsulated by the nonsolvent addition method include antibacterial and anticancer agents,[51] other pharmaceuticals,[52] steroids,[50] vitamins,[55] antacids,[56] paper-imaging dyes,[48,49] herbicides, and fertilizers.[49]

In general, the core material must be insoluble in both the solvent for the wall-forming polymer and the nonsolvent used for inducing polymer phase separation. In some cases, this requirement may be circumvented by modification of the core material to yield an insoluble form. For example, the soluble thioridazine free base was converted to its pamoate salt, which is insoluble in both the polymer solvent and the nonsolvent.[52]

Another means of microencapsulating a soluble core material is to initiate phase separation before introducing the core particles. This was illustrated in the microencapsulation of methylprednisolone, which is soluble in an ethanolic solution of the wall-forming styrene-maleic acid copolymer.[50] The nonsolvent isopropyl ether was added to the polymer solution until the first persistent turbidity appeared. The methylprednisolone particles were then added and the addition of nonsolvent was continued to complete the microencapsulation.

A special case in which a solvent-soluble core material can be microencapsulated was demonstrated by Vassiliades.[57] This procedure requires dissolving the core material in a water-insoluble oil and selecting a water-miscible solvent. The latter is a solvent for both a water-insoluble polymer and the oil containing the core material. The basis of this process is that the oil containing the core material will precipitate before the capsule wall material upon addition to an aqueous nonsolvent. In the patent example provided, a homogeneous solution was prepared by combining a solution of a leuko dye dissolved in soybean oil with a solution of polyvinyl chloride dissolved in tetrahydrofuran. When this combined solution was added to water, the oil containing the dye precipitated first and was encapsulated by the polymer during phase separation.

6. Double Encapsulation

The method of phase separation by nonsolvent addition has also been employed to prepare reservoir-type microcapsules using preformed core-polymer composites as the internal phase. In one example provided by Itoh and Nakano,[58] drug-cellulose acetate matrix particles were suspended in a solution of ethyl cellulose in diethyl ether. The matrix particles were previously obtained by dissolving cellulose acetate and the drug in acetone, evaporating the solvent, and milling the resultant sheet into particles. The addition of pentane induced the phase separation of ethyl cellulose which coated the suspended matrix particles to form the microcapsules.

Double-encapsulated microcapsules were prepared by Fong[52] using the low-temperature microencapsulation process in two consecutive operations. The process was used first to prepare the preformed microcapsules. In the first part of the procedure, thioridazine pamoate particles were dispersed in a solution of DL-polylactic acid in toluene. The polymer solution was previously cooled in a dry-ice bath. The addition of isopropanol as the nonsolvent induced phase separation of polylactic acid to form microcapsules. In the second part of the procedure, the above-mentioned preformed microcapsules were dispersed in a second polylactic acid solution which was previously cooled in a dry-ice bath. Isopropanol was added to induce phase separation of the polymer which deposited a second coating on the dispersed microcapsules.

Dissolution studies were conducted on the original and the double-encapsulated microcapsules. The initial release of the drug was found to be significantly reduced by the second polymeric coating.

Double-encapsulated microcapsules were also prepared in a two-part process patented by Hiestand.[49] The unique feature of this process is that a water-soluble core material can be microencapsulated with a water-soluble polymer. The preformed microcapsules were first fabricated with a water-insoluble polymer by the nonsolvent addition method. These microcapsules were then encapsulated with a water-soluble polymer by an aqueous phase separation process. Thus, these double-encapsulated microcapsules were comprised of two layers, an inner water-insoluble polymeric matrix encased by an outer water-soluble polymeric coating.

Another type of double-encapsulated microcapsule was prepared in a different two-part process patented by Reyes.[59] These microcapsules are the inverse of those fabricated by the above process of Hiestand[49] in that the inner core is derived from a water-soluble polymer and the outer coating is a water-insoluble polymer.

In the first step of this process, two w/o emulsions were prepared. The aqueous phase of one emulsion contained the core material in a solution of a gellable water-soluble polymer such as polyvinyl alcohol. The aqueous phase of the second emulsion contained borax, which is a gelling agent for polyvinyl alcohol. The organic phase of both emulsions contained a solution of water-insoluble polymer such as ethyl cellulose in xylene-carbon tetrachloride. Mixing the two emulsions produced a unified emulsion of gelled polyvinyl alcohol particles (containing the core material) dispersed in the ethyl cellulose solution.

The addition of petroleum ether in the second step effected deposition of the phase-separated ethyl cellulose on the dispersed gelled particles to form the double-encapsulated microcapsules.

C. Phase Separation Induced by Incompatible Polymer Addition

This process is similar to the previous one involving the addition of a nonsolvent. The procedure utilizes the addition of a polymer incompatible with the capsule wall material to induce phase separation of the latter. Addition of the second polymer causes the wall-forming polymer to be desolvated and to separate into a viscous, liquid phase. In a sense, the second polymer is acting as a nonsolvent for the capsule wall polymer.

Two separate phases are thus formed. One is the coacervated phase containing droplets of the wall-forming polymer. The other forms the continuous phase consisting of a solution of the incompatible polymer. The role of the incompatible polymer solution is not only to induce phase separation of the wall-forming polymer but also to stabilize it.

The core material is then encapsulated by droplets of the wall-forming polymeric phase to form the embryonic microcapsules. The newly formed microcapsules are usually too soft to isolate due to the presence of the polymer solvent. Hardening of the capsule wall is required before the microcapsules can be collected as discrete particles without agglomeration.

1. Incompatible Polymer

The incompatible polymer is selected so that it has a higher solubility in the solvent than the wall-forming polymer. Therefore, it is the wall-forming polymer which will phase separate out of solution. The second polymeric material must have little or no affinity for the core material so that the coacervate containing the wall-forming material will be the one which will preferentially encapsulate the core material.

Substances which have been used to induce phase separation include low molecular weight liquid polymers such as polybutadiene (8000 to 10,000 molecular weight), polydimethylsiloxane (500 cSt), and methacrylic polymer (325 cSt).[60] Phase separation can also be induced by addition of a nonpolymeric material such as light liquid paraffin[61] or vegetable oils.[62]

The advantage of using the incompatible polymer addition method is that the viscosity and relative volume of the coacervated phase can be controlled by the amount of incompatible polymer added.[58] This ability to control the viscosity of the coacervated phase could provide for better encapsulation of the core material than the nonsolvent addition method discussed in Section IV.B. In some cases involving the nonsolvent addition method, the viscosity of the coacervated phase may be too high or the capsule wall may harden too fast for efficient coating of the core material. Additional discussion on incompatible liquid polymeric agents is presented in Section IV.D.

2. Wall-Forming Polymer

Examples of polymers which have been used to form the capsule wall include ethyl cellulose, cellulose nitrate, polymethyl methacrylate, cellulose acetate phthalate, poly (acrylonitrile-co-styrene), polystyrene, and poly(acrylonitrile-co-vinylidene chloride).[60,61,63] An example of a biodegradable polymer which has received recent attention in this process is poly(lactide-co-glycolide).[62,64]

3. Hardening of Capsule Wall

Hardening of the microcapsule wall is usually done by adding an excess of a nonsolvent for the wall-forming polymer. The microcapsules can also be solidified by chemically cross-linking the capsule wall material or by solvent evaporation. Examples of these three means for hardening the capsule wall are outlined below.

Nonsolvent addition — A dispersion was formed by adding 4 g of methylene blue hydrochloride to a solution of 1 g of ethyl cellulose dissolved in 50 g of toluene.[63] The addition of 25 g of liquid polybutadiene induced phase separation of the ethyl cellulose which then encapsulated the dispersed core particles. The microcapsules were hardened by adding an excess of hexane, filtered, and dried.

Chemical cross-linking — A dispersion was formed by adding 4 g of ammonium nitrate to a solution of 1 g of ethyl cellulose dissolved in 50 g of 80:20 toluene-ethanol.[60] Phase separation of ethyl cellulose was induced by the addition of 25 g of liquid polybutadiene. The microcapsules were then hardened by adding 15 g of toluene diisocyanate which cross-linked with the residual ethanol in the capsule wall.

Solvent evaporation — In this example, phase separation was initiated by adding 350 mℓ of light liquid paraffin (Ondina 17®, manufactured by Shell Chemicals, Australia) to a solution of 4 to 10 g of ethyl cellulose dissolved in 600 mℓ of ethyl acetate.[61] After approximately 200 mℓ of solvent was evaporated by rapid agitation at 1500 rpm, 40 g of aspirin was introduced. Another third of the ethyl acetate was evaporated and 2000 mℓ of light liquid paraffin was added to complete phase separation of ethyl cellulose. The microcapsules were completely hardened after the remaining polymer solvent was removed by evaporation. Soft gelatin capsules were filled with the resulting slurry of microcapsules.

4. Control of Agglomeration

In the above example in which the capsule wall was hardened by solvent evaporation, agglomeration of microcapsules into large aggregates commonly occurred. This aggregation could be minimized by adding a large excess of light liquid paraffin immediately before agglomeration takes place. The optimum time for this addition was determined by trial runs.[61]

The problem of agglomeration may also be circumvented by employing the low-temperature microencapsulation process[52] described in Section IV.B.2.

5. Polymer Solvent

The solvent is selected to dissolve both the wall-forming polymer and the incompatible polymer employed for inducing phase separation. The polymer solvent must also be miscible with the nonsolvent if the latter is used to harden the capsule wall. However, it must not dissolve the core material.

Examples of polymer solvent include toluene, toluene-ethanol, ethyl acetate, methylene chloride, and ethylene chloride.

6. Nonsolvent

The nonsolvent for the wall-forming polymer has two functions. The first function is to harden the capsule wall and the other is to remove residual amounts of the incompatible polymer. Therefore, the nonsolvent must dissolve the incompatible polymer but not the wall-forming polymer or the core material. Aliphatic hydrocarbons, such as hexane or heptane, have been commonly used as the nonsolvent.

7. Core Material

Representative examples of core material microencapsulated by the incompatible polymer addition method include antibiotics,[64] other pharmaceuticals,[61] polypeptides,[62] and dyes.[63]

One technique for microencapsulating a soluble core material is to initiate phase separation before adding the core particles. An illustration of this was provided in the example in which the capsule wall was hardened by solvent evaporation (see Section IV.C.3). The aspirin particles were introduced for encapsulation after a portion of the polymer solvent was removed by evaporation.[61]

D. Phase Separation Induced by Temperature Change

This method takes advantage of the fact that some polymers are soluble in a hot solvent but become insoluble on cooling to room temperature. Therefore, temperature can be used as another means to induce phase separation of these wall-forming polymers. Accordingly, when such a polymer solution is allowed to cool, the wall-forming material becomes de-solvated and forms a separate polymer-rich phase. This viscous liquid phase can engulf the dispersed core material to form microcapsules. The capsule wall is then solidified upon cooling to ambient temperature.

This method was described in a patent issued to Fanger et al.[65] In an example, 8 g of magnesium hydride particles was stirred into a solution of 2 g of ethyl cellulose previously dissolved in 98 g of boiling cyclohexane. With continued stirring, the temperature of the mixture was allowed to cool slowly to room temperature. The core particles were encapsulated by the polymer-rich droplets which formed during this period of cooling. However, the microcapsules obtained were aggregated together into small, visible lumps. This aggregation is believed to be due to the tailing deposition of the wall-forming polymer.

Deasey et al.[66] indicated that careful attention to details of the procedure was necessary to minimize obtaining coarse, aggregated microcapsules. These precautions included vigorous agitation, slow cooling rate around the phase separation temperature, and washing the product with cold (10°C) cyclohexane.

1. Wall-Forming Polymer

The ideal wall polymer for this method should have substantially no solubility in a suitable solvent at room temperature. As the temperature of the solvent is increased above a certain temperature, the wall material should be completely dissolved. When the system is cooled below this critical phase separation temperature, the wall-forming polymer should separate as a polymer-rich liquid phase to encapsulate the dispersed core particles.

This method of inducing phase separation is limited to the few polymers which possess a high differential solubility with respect to temperature. The wall-forming polymer most commonly used in this process is ethyl cellulose dissolved in hot cyclohexane. Other polymers with this property include ethyl hydroxyethyl cellulose, methylcellulose, and hydroxypropyl methylcellulose.[65,77]

a. Polymer Molecular Weight

The molecular weight (grade) of ethyl cellulose used for preparing the capsule wall can affect the quality of the microcapsules. Polymer having a high ethoxy content (40 to 50%)

is preferred and is commercially available in a range of viscosities reflecting different molecular weights.

Deasey et al.[66] reported that 100-cP-grade (higher molecular weight) ethyl cellulose produced finer microcapsules with slower drug dissolution than those prepared with the 10-cP-grade (lower molecular weight) polymer. In both cases, more than 90% of the encapsulated sodium salicylate was released within 3 hr. This indicates that ethyl cellulose has a low capacity to retard dissolution of water-soluble drugs from microcapsules. Similar rapid release of water-soluble drugs from ethyl cellulose microcapsules was also reported by others.[67-75]

The effect of the molecular weight of ethyl cellulose on microencapsulation was also studied by Koida et al.[68] They found that aggregation of the microcapsules decreased to a minimum when the molecular weight was increased to approximately 15,000 to 20,000. Also, the release rate of encapsulated ascorbic acid depended significantly on the polymer molecular weight, with the slowest release obtained at a molecular weight in the range of 12,000 to 15,000.

b. Oleaginous Wall Material

The method of phase separation induced by temperature change is not limited to polymeric wall material. Some oleaginous materials were employed by Kondo and Nakano[54] to form the capsule wall. Examples of these materials are hydrogenated castor oil, hydrogenated beef tallow, cacao butter, and glycerin monostearate.

Unlike the polymeric materials, the coacervation of a hardened oil or fat usually occurs at a temperature much lower than the temperature at which it was dissolved. This coacervation process can continue for an extended period of time even at room temperature. Therefore, it is necessary that the system be stirred at the final temperature for a longer period of time to allow for sufficient phase separation.

2. Control of Agglomeration

In the beginning of Section IV.D, it was indicated that agglomeration of microcapsules could be minimized by employing vigorous agitation, slow cooling rate around the phase separation temperature, and washing the product with cold (10°C) cyclohexane.[66] Previously, Morse et al.[76] reported an improved procedure to prevent aggregation of the microcapsules prepared by this method. Since the wall of the microcapsules after phase separation is still soft due to solvation with cyclohexane, the microcapsules tend to aggregate during the drying step. By cooling to 10°C and washing the filtered microcapsules with pentane, hexane, heptane, or octane to displace the cyclohexane solvent, discrete particles were obtained without aggregation after drying.

In another patent, Morse et al.[77] reported that the core material can be added to the system after phase separation of the coating polymer has occurred. An advantage of this modified method is greater control of the process. In one example, a solution of 216 g of ethyl cellulose dissolved in 5 ℓ of cyclohexane was prepared by heating at 80°C. After cooling to 50°C, 1088 g of riboflavin particles was added. At this temperature, the phase-separated polymeric droplets were still sufficiently fluid to encapsulate the core particles. The temperature was raised to 57°C to complete the encapsulation. As in the previous patent,[76] the mixture was then cooled to 10°C to harden the capsule wall, filtered, washed twice with 1.5 ℓ of hexane, and dried. When the same process was conducted with both riboflavin particles and ethyl cellulose solution heated together at 80°C, the dispersion became very viscous. Large, aggregated microcapsules were obtained.

As previously indicated in Section IV.D.1.a, aggregation of the microcapsules was decreased with ethyl cellulose of a higher molecular weight.[66,68]

Donbrow and Benita[78] found that aggregation of the microcapsules was prevented by the inclusion of polyisobutylene in the process. In the absence of polyisobutylene, the encapsulated salicylamide waa an aggregated mass. In another paper,[79] they reported that poly-isobutylene was not coprecipitated but was adsorbed on the surface of the ethyl cellulose droplets produced by phase separation. Thus, polyisobutylene acted as a protective colloid in this microencapsulation process.

Samejima et al.[67] also reported that the presence of polyisobutylene prevented aggregation of the microcapsules. In addition, they evaluated the effects of butyl rubber and polyethylene on the microencapsulation of ascorbic acid. Although butyl rubber was equally effective as polyisobutylene in preventing aggregation, the release rate of ascorbic acid was the same as the nonencapsulated core material. Polyethylene was much less effective than polyiso-butylene in preventing aggregation.

The use of both butyl rubber and polyethylene for reducing agglomeration of microcapsules in these processes was previously reported in two patents.[80,81] However, Agyilirah and Nixon[75] reported that the presence of polyisobutylene increased rather than reduced aggre-gation of microcapsules containing phenobarbitone sodium.

3. Polymer Solvent

The solvent for the wall material is selected so that it is a good solvent for dissolving the polymer above the critical temperature but is a poor solvent below that temperature. Cy-clohexane is the preferred solvent for commonly used ethyl cellulose. For ethyl hydroxyethyl cellulose, a mixture of hydrocarbon solvents was used to dissolve the polymer.[65] This solvent mixture consisted of equal amounts of Shell Sol 140® and Shell Sol 72® (both are petroleum distillates but with different levels of paraffins). Ethanol was the preferred solvent for the oleaginous wall-forming material in the process of Kondo and Nakano.[54]

4. Core Material

Representative core materials which have been microencapsulated by the temperature-change method include sodium salicylate,[66] ascorbic acid,[67,68] chloramphenicol,[69] isoni-azid,[70,71] phenethicillin potassium,[72] phenobarbitone sodium,[73-75] niacinamide,[76] riboflavin,[77] and aspirin.[80,81]

Candidates for microencapsulation by this method must be insoluble in the polymer solvent at the phase separation temperature of the wall-forming polymer.

Another important prerequisite is that the core material be stable or stabilized at this temperature of about 80°C for a period of time. For example, to inhibit the hydrolysis of aspirin during encapsulation, a small amount of acetic anhydride was included in the fab-rication step.[80] For additional protection, the aspirin particles were also pretreated with a solution of an acid-buffering salt to retard hydrolysis in its encapsulated form during storage. Monobasic acid buffering salts which were used include sodium phosphate, potassium phos-phate, and ammonium phosphate.[81]

5. Double Encapsulation

This method was utilized by Suryakusuma and Jun[82] to apply a coating of ethyl cellulose around microspheres previously prepared by another procedure (suspension polymerization). A w/o dispersion was prepared by adding a hot solution of ethyl cellulose dissolved in cyclohexane to an aqueous solution of 2-hydroxyethyl methacrylate and acrylamide as the monomers, amaranth as the core material, and sodium persulfate as the polymerization initiator. The dispersion was maintained at 80°C with vigorous agitation for 1 hr. During this suspension-polymerization step, the ethyl cellulose acted as a protective colloid to prevent coalescence of the aqueous droplets by forming a protective film around these droplets. The dispersion was then allowed to cool, causing the ethyl cellulose coating to harden as a membrane around the polymer beads.

E. Phase Separation Induced by Solvent Evaporation

Other solvent evaporation processes described elsewhere in this chapter involve systems consisting of either an aqueous (Section III) or an organic emulsion (two immiscible liquid phases, Section IV.C.3). This process is different in that it involves solvent removal from a system in which the core material is dispersed in a liquid phase which is initially homogeneous. This liquid phase consists of the wall-forming polymer solution and a suspending liquid. The polymer solvent is selected such that it is completely miscible with and has a boiling point lower than the suspending liquid. The suspending liquid is a nonsolvent for the wall-forming polymer. The relative amounts of the wall-forming polymer, polymer solvent, and the suspending liquid are such that the resulting solution exhibits no initial phase separation. As the system is heated, the polymer solvent is gradually evaporated. This causes the wall-forming polymer to separate into viscous liquid droplets which become dispersed in the suspending liquid. Microcapsules are formed when the core material is coated by these polymer-rich droplets.

In an example of this method described by Yoshida,[83] 36 g of potassium dichromate particles was dispersed in a homogeneous solution consisting of 4 g of ethyl cellulose as the capsule wall material, 80 g of acetone as the polymer solvent, and 140 g of a relatively nonvolatile hydrocarbon liquid (Dispersol-81515®) as the suspending liquid. The dispersion was then heated at about 50°C. During the evaporation of the acetone, a viscous liquid phase rich in ethyl cellulose separated and enveloped the core particles to form microcapsules. When all of the solvent had been evaporated, the microcapsules were collected and washed with a volatile petroleum distillate to remove residual suspending liquid.

Examples of suspending liquids include high boiling point hydrocarbons, fluorinated hydrocarbons, silicone fluids and oils, and polyethylene glycols.

Besides ethyl cellulose, examples of other wall-forming polymers include nitrocellulose, polyethylene, ethylene-acrylic acid copolymers, vinylidene chloride polymers, chlorinated natural rubber, and vinyl polymers.

Polymer solvents which can be used are volatile aliphatic and aromatic hydrocarbons, ketones, ethers, and alcohols.

REFERENCES

1. **Bissery, M.-C., Thies, C., and Puisieux, F.,** A study of process parameters in the making of microspheres by the solvent evaporation process, in Proc. 10th Int. Symp. Controlled Release of Bioactive Materials, San Francisco, July 1983, 262.
2. **Thies, C. and Bissery, M.-C.,** Biodegradable microspheres for parenteral administration, in *Biomedical Applications of Microencapsulation,* Lim, F., Ed., CRC Press, Boca Raton, Fla., 1984, 53.
3. **Fong, J. W., Nazareno, J. P., Pearson, J. E., and Maulding, H. V.,** Evaluation of biodegradable microspheres prepared by a solvent evaporation process using sodium oleate as emulsifier, *J. Controlled Release,* 3, 119, 1986.
4. **Bissery, M.-C., Puisieux, F., and Thies, C.,** Preparation and characterization of poly (β-hydroxybutyrate) microspheres, in Proc. 9th Int. Symp. Controlled Release of Bioactive Materials, Fort Lauderdale, Fla., July 1982, 30.
5. **Courteille, F., Lenk, T., and Thies, C.,** The structure and aging of progesterone-loaded microspheres, in Proc. 11th Int. Symp. Controlled Release of Bioactive Materials, Fort Lauderdale, Fla., July 1984, 90.
6. **Cardinal, J. R.,** Matrix systems, in *Medical Applications of Controlled Release,* Vol. 1, Langer, R. S. and Wise, D. L., Eds., CRC Press, Boca Raton, Fla., 1984, 41.
7. **Deasey, P. B.,** *Microencapsulation and Related Drug Processes,* Marcel Dekker, New York, 1984, chap. 14.

8. **Good, W. R. and Lee, P. I.,** Membrane controlled reservoir drug delivery systems, in *Medical Applications of Controlled Release,* Vol. 1, Langer, R. S. and Wise, D. L., Eds., CRC Press, Boca Raton, Fla., 1984, 1.

9. **Peppas, N. A.,** Mathematical models for controlled release kinetics, in *Medical Applications of Controlled Release,* Vol. 2, Langer, R. S. and Wise, D. L., Eds., CRC Press, Boca Raton, Fla., 1984, 169.

10. **Roseman, T. J.,** Monolithic polymer devices — Section 1, in *Controlled Release Technologies: Methods, Theory, and Applications,* Vol. 1, Kydonieus, A. F., Ed., CRC Press, Boca Raton, Fla., 1980, 21.

11. **Visscher, G. E., Robison, R. L., Maulding, H. V., Fong, J. W., Pearson, J. E., and Argentieri, G. J.,** Biodegradation of and tissue reaction to 50:50 poly(DL-lactide-co-glycolide) microcapsules, *J. Biomed. Mater. Res.,* 19, 349, 1985.

12. **Visscher, G. E., Robison, R. L., Maulding, H. V., Fong, J. W., Pearson, J. E., and Argentieri, G. J.,** Note: Biodegradation of and tissue reaction to poly(DL-lactide) microcapsules, *J. Biomed. Mater. Res.,* 20, 667, 1986.

13. **Rosen, M. J.,** *Surfactants and Interfacial Phenomena,* John Wiley & Sons, New York, 1978, 242.

14. **Fong, J. W.,** Process for Preparation of Microspheres, U.S. Patent 4,384,975, 1983.

15. **Fong, J. W.,** Process for Preparation of Microspheres and Modification of Release Rate of Core Material, U.S. Patent 4,479,911, 1984.

16. **Fong, J. W., Maulding, H. V., Visscher, G. E., Nazareno, J. P., and Pearson, J. E.,** Enhancing drug release from polylactide microspheres by using base in the microcapsulation process, in *Controlled-Release Technology: Pharmaceutical Applications,* Lee, P. I. and Good, W. R., Eds., American Chemical Society, Washington, D.C., 1987, chap. 16.

17. **Morishita, M., Inaba, Y., Fukushima, M., Hattori, Y., Kobari, S., and Matsuda, T.,** Process for Encapsulation of Medicaments, U.S. Patent 3,960,757, 1976.

18. **Wakiyama, N., Juni, K., and Nakano, M.,** Preparation and evaluation *in vitro* and *in vivo* of polylactic acid microspheres containing dibucaine, *Chem. Pharm. Bull.,* 30, 3719, 1982.

19. **Jaffe, H.,** Microencapsulation Process, U.S. Patent 4,272,398, 1981.

20. **Wakiyama, N., Juni, K., and Nakano, M.,** Influence of physicochemical properties of polylactic acid on the characteristics and *in vitro* release patterns of polylactic acid microspheres containing local anesthetics, *Chem. Pharm. Bull.,* 30, 2621, 1982.

21. **Yolles, S., Leafe, T., Sartori, M., Torkelson, M., and Ward, L.,** Controlled release of biologically active agents, in *Controlled Release Polymeric Formulations,* Paul, D. R. and Harris, F. W., Eds., American Chemical Socity, Washington, D.C., 1976, 123.

22. **Beck, L. R., Cowsar, D. R., Lewis, D. H., Cosgrove, R. J., Jr., Riddle, C. T., Lowry, S. L., and Epperly, T.,** A new long-acting injectable microcapsule system for the administration of progesterone, *Fertil. Steril.,* 31, 545, 1979.

23. **Wakiyama, N., Juni, K., and Nakano, M.,** Preparation and evaluation *in vitro* of polylactic acid microspheres containing local anesthetics, *Chem. Pharm. Bull.,* 29, 3363, 1981.

24. **Van Besauw, J. F., and Claeys, D. A.,** Method for Encapsulating Aqueous or Hydrophilic Material, U.S. Patent 3,645,911, 1972.

25. **Beck, L. R., Pope, V. Z., Flowers, C. E., Jr., Cowsar, D. R., Tice, T. R., Lewis, D. H., Dunn, R. L., Moore, A. B., and Gilley, R. M.,** Poly(DL-lactide-co-glycolide)/norethisterone microcapsules: an injectable biodegradable contraceptive, *Biol. Reprod.,* 28, 186, 1983.

26. **Setterstrom, J. A., Tice, T. R., Lewis, D. H., and Meyers, W. E.,** Controlled release of antibiotics from biodegradable microcapsules for wound infection control, in Proc. 1982 Army Science Conf., West Point, June 1982, 215.

27. **Bissery, M.-C., Valeriote, F., and Thies, C.,** In vitro lomustine release from small poly(β-hydroxybutyrate) and poly(D,L-lactide) microspheres, in Proc. 11th Int. Symp. Controlled Release of Bioactive Materials, Fort Lauderdale, Fla., July 1984, 25.

28. **Goosen, T., O'Shea, G., Chou, S., and Sun, A.,** Slow release of insulin from biodegradable microbeads injected in diabetic rats, in Proc. 10th Int. Symp. Controlled Release of Bioactive Materials, San Francisco, July 1983, 274.

29. **Morishita, M., Fukushima, M., and Inaba, Y.,** Microencapsulation of activated charcoal and its biochemical applications, in *Microencapsulation: Processes and Applications,* Vandegaer, J. E., Ed., Plenum Press, New York, 1974, 115.

30. **Ishibashi, K., Noda, Y., Yamada, K., Nozawa, Y., Miyagishima, A., and Higashide, F.,** Preparation of microcapsules containing activated carbon, in Proc. 10th Int. Symp. Controlled Release of Bioactive Materials, San Francisco, July 1983, 268.

31. **Jaffe, H., Giang, P. A., and Miller, J. A.,** Implants of methoprene in poly(lactic acid) against cattle grubs, in Proc. 5th Int. Symp. Controlled Release of Bioactive Materials, Gaithersburg, Md., August 1978, 55.

32. **Tice, T. R., Lewis, D. H., and Cowsar, D. R.,** Controlled release of pesticide from chlorpyrifos micro-capsules, in Proc. 8th Int. Symp. Controlled Release of Bioactive Materials, Fort Lauderdale, Fla., July 1981, 135.

33. **Coleman, D. R., Mason, D. W., Dappert, T. O., Tice, T. R., and Meyers, W. E.,** Detection of toxic compounds with microencapsulated reagents, in Proc. 11th Int. Symp. Controlled Release of Bioactive Materials, Fort Lauderdale, Fla., July 1984, 75.

34. **Maulding, H. V., Tice, T. R., Cowsar, D. R., Fong, J. W., Pearson, J. E., and Nazareno, J. P.,** Biodegradable microcapsules: acceleration of polymeric excipient hydrolytic rate by incorporation of a basic medicament, *J. Controlled Release,* 3, 103, 1986.

35. **Tice, T. R. and Lewis, D. H.,** Microencapsulation of Pharmaceuticals, U.S. Patent 4,389,330, 1983.

36. **Vrancken, M. N. and Claeys, D. A.,** Process for Encapsulating Water and Compounds in Aqueous Phase by Evaporation, U.S. Patent 3,523,906, 1970.

37. **Kondo, A.,** Microencapsulation utilizing in-liquid drying process (complex emulsion method), in *Micro-capsule Processing and Technology,* Van Valkenburg, J. W., Ed., Marcel Dekker, New York, 1979, 106.

38. **Vrancken, M. N. and Claeys, D. A.,** Method for Encapsulating Water and Compounds in Aqueous Phase by Extraction, U.S. Patent 3,523,907, 1970.

39. **Kruse, U. and Herman, D. F.,** Finely Divided Hollow Microcapsules of Polymeric Resins, U.S. Patent 3,784,391, 1974.

40. **Gilley, R. M., Tice, T. R., Staas, J. K., and Beck, L. R.,** Development of controlled-release progesterone microcapsules for the regulation of fertility during lactation, in Proc. 11th Int. Symp. Controlled Release of Bioactive Materials, Fort Lauderdale, Fla., July 1984, 73.

41. **Beck, L. R. and Tice, T. R.,** Poly(lactic acid) and poly(lactic acid-co-glycolic acid) contraceptive delivery systems, in *Long-Acting Steroid Contraception,* Mishell, D. R., Jr., Ed., Raven Press, New York, 1983, 175.

42. **Pitt, C. G., Jeffcoat, A. R., Zweidinger, R. A., and Schindler, A.,** Sustained drug delivery systems. I. The permeability of poly (ϵ-caprolactone), poly(DL-lactic acid), and their copolymers, *J. Biomed. Mater. Res.,* 13, 497, 1979.

43. **Pitt, C. G., Gratzl, M. M., Jeffcoat, A. R., Zweidinger, R., and Schindler, A.,** Sustained drug delivery systems. II. Factors affecting release rates from poly(ϵ-caprolactone) and related biodegradable polyesters, *J. Pharm. Sci.,* 68, 1534, 1979.

44. **Schindler, A., Jeffcoat, R., Kimmel, G. L., Pitt, C. G., Wall, M. E., and Zweidinger, R.,** Biode-gradable polymers for sustained drug delivery, in *Contemporary Topics in Polymer Science,* Vol. 2, Pearce, E. M. and Schaefgen, J. R., Eds., Plenum Press, New York, 1977, 251.

45. **Pitt, C. G. and Schindler, A.,** Biodegradation of polymers, in *Controlled Drug Delivery, Vol. 1,* Bruck, S. D., Ed., CRC Press, Boca Raton, Fla., 1983, 53.

46. **Tice, T. R., Lewis, D. H., Dunn, R. L., Meyers, W. E., Casper, R. A., and Cowsar, D. R.,** Biodegradation of microcapsules and biomedical devices prepared with resorbable polyesters, in Proc. 9th Int. Symp. Controlled Release of Bioactive Materials, Fort Lauderdale, Fla., July 1982, 21.

47. **Nozawa, Y. and Higashide, F.,** Drug release from and properties of poly(styrene) microcapsules, in *Polymeric Delivery Systems,* Kostelnik, R. J., Ed., Gordon and Breach Science, New York, 1978, 101.

48. **Reyes, Z.,** Polymeric Microcapsules for Coatings, U.S. Patent 3,173,878, 1965.

49. **Hiestand, E. N.,** Coating by Phase Separation, U.S. Patent 3,242,051, 1966.

50. **Rowe, E. L.,** Process of Coating Particles with a Polymer, U.S. Patent 3,336,155, 1967.

51. **Itoh, M., Nakano, M., Juni, K., and Sekikawa, H.,** Sustained release of sulfamethizole, fluorouacil, and doxorubicin from ethylcellulose-polylactic acid microcapsules, *Chem. Pharm. Bull.,* 28, 1051, 1980.

52. **Fong, J. W.,** Processes for Preparation of Microspheres, U.S. Patent 4,166,800, 1979.

53. **Vandegaer, J. E. and Meier, F. C.,** Encapsulation Process, U.S. Patent 3,575,882, 1971.

54. **Kondo, S. and Nakano, M.,** Microcapsules, U.S. Patent 4,102,806, 1978.

55. **El-Sayed, A. A., Badawi, A. A., and Fouli, A. M.,** Effect of solvent used in the preparation of solid dispersions and microcapsules on the dissolution of drugs, *Pharm. Acta Helv.,* 57, 61, 1982.

56. **Kasai, S. and Koishi, M.,** Studies on the preparation of ethylcellulose microcapsules containing magnesium aluminum hydroxide hydrate, *Chem. Pharm. Bull.,* 25, 314, 1977.

57. **Vassiliades, A. E.,** Microcapsules and Transfer-Sheet Record Material Coated Therewith, U.S. Patent 3,418,250, 1968.

58. **Itoh, M. and Nakano, M.,** Sustained release of drugs from ethylcellulose microcapsules containing drugs dispersed in matrices, *Chem. Pharm. Bull.,* 28, 2816, 1980.

59. **Reyes, Z.,** Microcapsules, U.S. Patent 3,405,070, 1968.

60. **Powell, T. C., Steinle, M. E., and Yoncoskie, R. A.,** Process of Forming Minute Capsules En Masse, U.S. Patent 3,415,758, 1968.

61. **D'Onofrio, G. P., Oppenheim, R. C., and Bateman, N. E.,** Encapsulated microcapsules, *Int. J. Pharm.,* 2, 91, 1979.

62. **Kent, J. S., Sanders, L. M., Lewis, D. H., and Tice, T. R.,** Microencapsulation of Water Soluble Polypeptides, European Patent Application 52510, May 26, 1982.

63. National Cash Register, British Patent 907,284, 1963.

64. **Tice, T. R., Meyers, W. E., Lewis, D. H., and Cowsar, D. R.,** Controlled release of ampicillin and gentamicin from biodegradable microcapsules, in Proc. 8th Int. Symp. Controlled Release of Bioactive Materials, Fort Lauderdale, Fla., July 1981, 108.

65. **Fanger, G. O., Miller, R. E., and McNiff, R. G.,** En Masse Encapsulation Process, U.S. Patent 3,531,418, 1970.

66. **Deasey, P. B., Brophy, M. R., Ecanow, B., and Joy, M. M.,** Effect of ethylcellulose grade and sealant treatments on the production and in vitro release of microencapsulated sodium salicylate, *J. Pharm. Pharmacol.,* 32, 15, 1980.

67. **Samejima, M., Hirata, G., and Koida, Y.,** Studies on microcapsules. I. Role and effect of coacervation-inducing agents in the microencapsulation of ascorbic acid by a phase separation method, *Chem. Pharm. Bull.,* 30, 2894, 1982.

68. **Koida, Y., Hirata, G., and Samejima, M.,** Studies on microcapsules. II. Influence of molecular weight of ethylcellulose in the microencapsulation of ascorbic acid, *Chem. Pharm. Bull.,* 31, 4476, 1983.

69. **Salib, N. N., El-Menshawy, M. E., and Ismail, A. A.,** Ethyl cellulose as a potential sustained release coating for oral pharmaceuticals, *Pharmazie,* 31, 721, 1976.

70. **Jalsenjak, I., Nixon, J. R., Senjkovic, R., and Stivic, I.,** Sustained-release dosage forms of microencapsulated isoniazid, *J. Pharm. Pharmacol.,* 32, 678, 1980.

71. **Senjkovic, R. and Jalsenjak, I.,** Effect of capsule size and membrane density on permeability of ethyl cellulose microcapsules, *Pharm. Acta Helv.,* 57, 16, 1982.

72. **Alpar, H. O. and Walters, V.,** The prolongation of the in vitro dissolution of a soluble drug (phenethicillin potassium) by microencapsulation with ethyl cellulose, *J. Pharm. Pharmacol.,* 33, 419, 1981.

73. **Jalsenjak, I., Nicolaidou, C. F., and Nixon, J. R.,** The *in vitro* dissolution of phenobarbitone sodium from ethyl cellulose microcapsules, *J. Pharm. Pharmacol.,* 28, 912, 1976.

74. **Nixon, J. R., Jalsenjak, I., Nicolaidou, C. F., and Harris, M.,** Release of drugs from suspended and tabletted microcapsules, *Drug Devel. Indust. Pharm.,* 4, 117, 1978.

75. **Agyilirah, G. A. and Nixon, J. R.,** Preparation, tabletting and release characteristic of ethyl cellulose-walled microcapsules of phenobarbitone sodium, *Acta Pharm. Technol.,* 26, 251, 1980.

76. **Morse, L. D., Boroshok, M. J., and Grabner, R. W.,** Isolating Cyclohexane-Free Ethylcellulose Microcapsules, U.S. Patent 4,107,072, 1978.

77. **Morse, L. D., Walker, W. G., and Hammes, P. A.,** Microencapsulation, U.S. Patent 4,123,382, 1978.

78. **Donbrow, M. and Benita, S.,** The effect of polyisobutylene on the coacervation of ethyl cellulose and the formation of microcapsules, *J. Pharm. Pharmacol.,* 29 (Suppl.), 4P, 1977.

79. **Benita, S. and Donbrow, M.,** Coacervation of ethyl cellulose: the role of polyisobutylene and the effect of its concentration, *J. Colloid Interface Sci.,* 77, 102, 1980.

80. **Miller, R. E. and Anderson, J. L.,** Encapsulation Process and its Product, U.S. Patent 3,155,590, 1964.

81. **Anderson, J. L., Gardner, G. L., and Yoshida, N. H.,** Encapsulation of Aspirin in Ethyl Cellulose and its Product, U.S. Patent 3,341,416, 1967.

82. **Suryakusuma, H. and Jun, H. W.,** Formation of encapsulated hydrophilic polymer beads by combined techniques of bead polymerization and phase separation, *J. Pharm. Pharmacol.,* 36, 493, 1984.

83. **Yoshida, N. H.,** Preparation of Minuscule Capsules, U.S. Patent 3,657,144, 1972.

Chapter 6

MECHANOCHEMICAL ENCAPSULATION PROCESS BY DRY BLENDING

Masumi Koishi and Takafumi Ishizaka

TABLE OF CONTENTS

I. INTRODUCTION

It is well known that fine cohesive particles easily adhere to the surface of a coarser excipient and this type of mixture can have pharmaceutical or cosmetic applications.

In the preparation of granules and tablets, the blending of pharmaceutical solids and excipients or disintegrators is done easily with equipment such as planetary mixers, nauta mixers, revolvo-cube mixers, V-type mixers, or automatic mortar. However, the conditions of ordered powder mixing, such as the degree of blending and the blendability of powders, play an important role in determining the physical and mechanical properties of granules and tablets after their preparation. Generally speaking, in the manufacture of solid dosage forms, the preferential adhesion of powders to the surfaces of other powders were observed to result in some cases from electrostatic charging, sticking, or friction of fine cohesive powders during blending. These frictional, charging, and physical adhesiveness properties can be used to prepare mechanically ordered mixtures or to modify and encapsulate the surface of drug solids.

Hersey[1] first described the formation of ordered mixes in which interparticle adhesion is responsible for the bonding of fine adherent particles of one constituent powder to coarse "carrier" particles of a second system. Studies carried out on various pharmaceutical ordered mixes showed that under certain conditions interparticle attraction is incapable of preventing adhered particles from being removed from the surface of carrier particles. This suggests that the overall stability of an ordered mix during processing and handling will be determined by the strength of interparticle attraction in ordered units. Interparticle attraction can be used practically in pharmaceutical preparations for the manufacture of powder-encapsulated pharmaceuticals. Coated or adhesional ordered mixtures are obtained through considerable care in blending procedures.

In more detailed discussions, mention has been made of the widespread use of a final treatment of drug-loaded potato starch or lactose with small (up to 3% by weight) amounts of various organic compounds such as carnauba wax, polyvinylpyrrolidone, carrageenan, glycerylmonostearate, and magnesium stearate (Mg-St). As in many other areas of coating, granulation, or microencapsulation, the amount of published information of the effects of these compounds on the surface character of drug-loaded potato starch or lactose is small, despite voluminous patent literature. Most patents on this subject claim that the addition of such compounds gives drugs an increased or decreased speed of wetting and improved disintegration or dispersion stability in special defined media. Thus, in determining the texture effect of such compounds on the surface character of the drug and therefore its behavior in acid or alkaline media, it is necessary to determine if the organic compound has any positive benefit in the actual oral administration of drugs or if the organic compound interferes with the interaction between the drug surface and acid or alkaline media. Thiel and Nguyen[2] describe results obtained by studying the fluidized bed granulation of a 0.1% ordered mixture. The mixture was granulated using a 5% aqueous solution of polyvinyl-pyrrolidone. From the fluidized bed granulation it is clear that a 0.1% ordered mixture of cohesive salicylic acid and spray-dried lactose excipient is stable when fluidized and no significant loss of salicylic acid occurs during fluidized bed granulation. Granulation of the ordered mixture fixes the ordered units in a random pattern. Salicylic acid is much more uniformly distributed in the granules than in the original ordered mixtures. Granulation offers a viable method of greatly reducing the effect of ordered unit segregation, thus eliminating the need to use closely sized carrier materials. From these results it can be seen that a fluidized bed is a very efficient mixer for randomizing non- or slightly cohesive materials, but the level of energy input may not be sufficient to break down agglomerates of highly cohesive particles. Dissolution and bioavailability requirements often indicate the use of very fine active ingredients; the processing method described is ideally suited to producing very homogeneous granules containing small quantities of active materials. The other application discussed by Thiel and Nguyen[3] was a fluidized bed film coating of an ordered mixture to produce microencapsulated ordered units. In this experiment, very finely divided particulate salicylic acid (2 to 5 μm) was microencapsulated using conventional fluidized-bed coating techniques. The process involved spray coating a preformed ordered mixture, in which the micronized particles are adsorbed on the surface of a coarser carrier material, with a 5% solution of cellulose acetate phthalate in aqueous sodium hydroxide (pH 7 to 8). The mixtures contained 0.1, 1.0, and 5.0% weight of a microfine model drug (salicylic acid). Between 75 and 95% of the micronized particles were retained on the carrier surface beneath the film. This process offers a novel method of microencapsulating very fine particulate materials, which may find an application in the production of enteric coated and sustained release microdose products.

On the other hand, Ampolsuk et al.[4] pointed out the influence of the dispersion method on the dissolution rate of digoxin-lactose and hydrocortisone-lactose triturations. Their report described the dissolution of digoxin and hydrocortisone from lactose triturations prepared by the following three methods: (1) simple blending, (2) solvent deposition, and (3) frictional pressure. The third method was included because steroidal compounds might be spreadable on the diluent surface by frictional force.

Ampolsuk et al. concluded, from the above experiments, that the spreading of digoxin or hydrocortisone over a lactose surface by frictional pressure produced 1:20 triturations with significantly enhanced dissolution rates in simulated gastric fluid. Frictional forces are applied to varying degrees in the manufacture of such dosage forms. Procedures such as milling, blending, slugging, granulating, and tableting impart frictional force on the drug and the excipients. The intensity and length of time for each operation will vary. In light of the reported results, it is not surprising that such procedural variations could account for the dissolution rate variations encountered. Triturations prepared by simple blending or solvent deposition exhibited slower dissolution rates. These observations may help to explain the variations in dissolution rate observed in manufactured tablet and capsule dosage forms of these two drugs.

II. MECHANOCHEMICAL ENCAPSULATION OF AN ORDERED POWDER MIXTURE AND ITS APPLICATIONS TO SUSTAINED RELEASE PREPARATIONS

A. Fundamental Considerations of Powder Mixing

The random mixing theory has been extensively explored.[5,6] Randomization requires equally sized and weighted particles with little or no surface effects, showing no cohesion or interparticle interaction, to achieve the best results. However, it cannot be applied to all practical mixing situations, especially where cohesive or interacting particles are mixed. Travers and White[7] showed that fine cohesive particles adhered on the surface of a coarser excipient. The term "ordered mixing" was given to this phenomenon by Hersey.[1] The concept of ordered mixing is useful in explaining powder mixing of cohesive or interacting fine particles. Ordered mixing, furthermore, is different from random mixing in that it does not require equally sized or weighted particles; it requires particle interaction, i.e., van der Waals forces, surface tension, frictional pressure, electrostatic charge, or any other form of adhesion.[8-13]

Ordered mixtures are frequently more homogeneous than random mixtures.[14] The standard deviations of ordered mixtures are unaffected by sample size. If complete ordered mixing is achieved, the standard deviation will be zero. The pharmaceutical applications of ordered mixing, which has potential advantages for the manufacture of low-dosage tablets and capsules, have been studied. Crooks and Ho[15] and Johnson[16] have investigated the use of

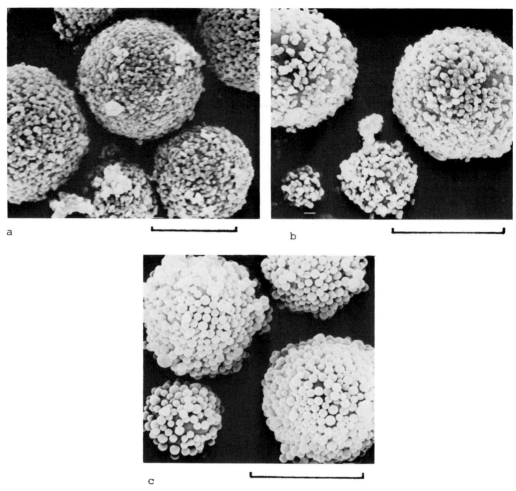

a

b

c

FIGURE 1.   Scanning electron micrographs of modified-plastic particles: (a) titanium dioxide (bare, rutile type) modified-nylon 12 particles, (b) titanium dioxide modified-polyethylene particles, (c) polymethylmethacrylate modified-polyethylene particles. Scale 5 = μm.

ordered mixture for direct compression tableting. Thiel and Nguyen[17] granulated ordered mixtures. Recently, the development of sustained drug delivery systems has been a major research area in the field of pharmaceutics. It has been reported that the addition of a hydrophobic lubricant such as Mg-St to solid dose formulations decreases the release rate of drug.[18-20] It has also been said that the crushing strength of tablets dramatically decreases as mixing time of particulate solids with Mg-St increases and that the phenomenon is caused by the formation of a lubricant film upon the substrate. This film is the result of the adhesion to the substrate surface of Mg-St molecules which are sheared off from the Mg-St crystals during the mixing process.[21-25]

B. Powder Mixing by Triboelectrification

The concept of ordered mixing, in which one component consisting of fine particles adheres to a second component of coarser carrier particles, has been proposed as a theoretically ideal system capable of producing near-perfect homogeneous mixes.[1] Such systems have previously been considered unsuitable for the production of homogeneous mixes because of the difference in particle size of the two components, which has been shown to produce segregation of the ingredients of a random powder mix. However, an almost-perfect homogeneous mix has been attempted, using triboelectrification[9,10] as a novel powder mixing process. Figure 1 shows typical scanning electron microscope (SEM) photographs of batch mixing 35% (w/w) titanium dioxide or polymethylmethacrylate and polyethylene or nylon 12 powder.

All modified powders were prepared by blending at room temperature for 30 min with an automatic ceramic mortar. The pestle rotated at 100 rpm and the mortar rotated inversely at 6 rpm in operation. In Figure 1, each small particle adhered to the surface of large core particles. The desired degree of hydrophilicity of products can be achieved by varying the weight ratio of both particles. From the mixing experiments, several observations on formation mechanism were made. For example, when particles of nylon 12 are blended with titanium dioxide particles in an automatic mortar, nylon 12 particles are not pulverized, but titanium dioxide adheres to the surface of the particles. During this treatment, friction among particles of nylon 12 and titanium dioxide causes the particles of nylon 12 to be electrically charged. By this frictional charge and the adhesive effect of a minor amount of free water adsorbed on the surfaces of nylon 12 particles from moisture in the air, hydrophilic titanium dioxide adheres to the surface of nylon 12 particles in the form of single particles or aggregates. The ratio of the inorganic pigment to the particles of the polymeric compound can be controlled by changing the mixing ratio of inorganic pigment to particles of the polymeric compound. In this way, hydrophilic and oleophilic portions are formed on the surfaces of particles and both hydrophilic and oleophilic drug solutions can be adsorbed on the particles. Figure 2 shows SEM photographs illustrating the states of a pollen grain of both *Lilium concolor* and *Elodea densa*.

C. Encapsulation of an Ordered Powder Mixture

1. Preparation of Ordered Mixtures and Encapsulation of Mixtures

Ordered mixing was carried out in a centrifugal rotating mixer (Mechanomill MM-10, Okada Seiko Co., Tokyo, Japan) equipped with four buffer plates, as shown in Figure 3. In the rotating vessel the powders are mixed by convection depending on both centrifugal force and the effect of buffer plates.

The vessel containing 0.1 g isoproterenol HCl, 9.9 g potato starch, and ten stainless steel balls (50 mm in diameter) was rotated at 300 rpm for 10 min in order to destroy powder agglomerates formed during the mixing. The operation was carried out at 23°C. Then, 0.3 g Mg.St was added to 10 g ordered mixture and mixed for 60 min at 35, 50, and 70°C under an infrared lamp. Mixing was carried out in the same way for all experiments. Dry potato starch (JP grade), an excipient, used as a carrier was sieved in the range of 200 to 250 mesh. Isoproterenol HCl (a drug for asthma, used as a model drug) and Mg-St (a hydrophobic lubricant, used as a wall material) were crushed to under 5 μm by Jet-mill.

After mixing, 20 200-mg samples were taken randomly from ordered mixtures and Mg-St-coated ordered mixtures, respectively. Each sample was added to 100 mℓ of 0.01 N HCl and heated up to 75°C. After the fatty acid salt became transparent, the mixture was cooled to room temperature, followed by the addition of 0.01 N HCl to adjust the total volume to 100 mℓ. It was then filtered and the light absorption of the filtrate was measured at 278 nm on a UV spectrophotometer. The isoproterenol HCl concentration in the filtrate was determined from the standard Beer's law plot. For 20 samples the degree of mixing was determined by the coefficient of variation (CV), a useful measure of segregation intensity, which was calculated by the following equation:

$$CV = \frac{\sigma}{\chi} \tag{1}$$

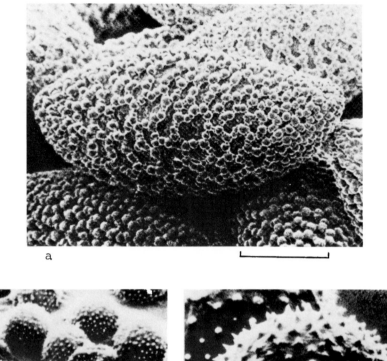

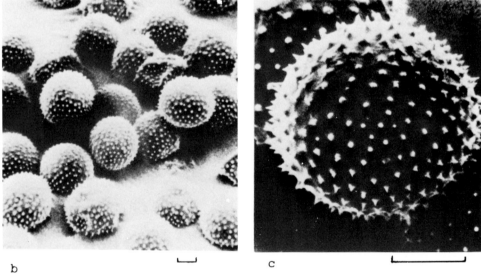

FIGURE 2.   Scanning electron micrographs of pollen grains: (a) *Lilium concolor* and (b and c) *Elodea densa*. Scale = 20 μm. (From Morisaki, M., in *Micrographic Plant Structures*, Ueda, R., Ed., Morikita, Tokyo, 1983, 177. With permission.)

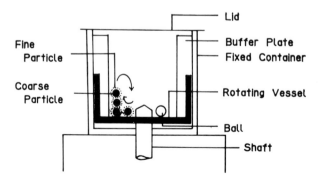

FIGURE 3.   Schematic diagram of centrifugal rotating mixer.

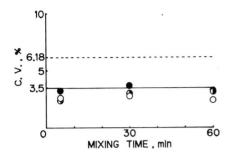

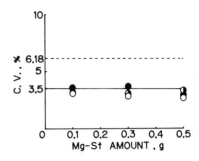

FIGURE 4.   CV for Mg-St-coated ordered mixtures as a function of mixing time: ○, prepared at 35°C; ◑, 50°C; and ●, 70°C. The amount of Mg-St is 0.3 g.

FIGURE 5.   CV for Mg-St-coated ordered mixtures as a function of Mg-St amount: ○, prepared at 35°C; ◑, 50°C; and ●, 70°C. Mixing time is 30 min.

where σ and χ are the standard deviation and mean, respectively.

2. Evaluation of Degree of Mixing

Kozatani et al.[26] showed that good miscibility was attained if the value of CV was less than 6.18%, which was obtained by assuming that the drug content gives a normal distribution and, for at least 90% of the mixtures, is within the $100 \pm 10\%$ region. Therefore, in the present study the degree of mixing was evaluated to be good when the value of CV averaged less than 6.18%. Figure 4 shows the effect of mixing time on the degree of mixing at three different temperatures. For any temperature or any mixing time, the CV value was less than 6.18% and almost equal to 3.5%, as can be seen in Figure 4. Figure 5 shows the effect of the amount of Mg-St on the degree of mixing at three different temperatures. The same trend was found as in the mixing time effect (see Figure 4). From these results, it is confirmed that mixing of drug and excipient was well done whether Mg-St was present or not.

3. Observations of Surface Appearance

Figures 6 to 9 show the Mg-St-coated ordered mixtures under various mixing conditions. Figure 6 shows the Mg-St-coated ordered mixtures and their surface appearances after adding 0.3 g Mg-St to the ordered mixtures and mixing for 60 min at 35, 50, and 70°C, respectively. Judging from the photos, all the Mg-St particles appear to have adhered to the ordered mixtures with no free salt particles at any temperature. Many fine Mg-St particles are seen on the surface in Figure 6 a-1 and a-2. The sample prepared at 35°C has the roughest surface of the three samples. The surface appearance of the Mg-St-coated ordered mixtures becomes increasingly smoother as the preparation temperature becomes higher. At 70°C in Figure 6 c-1 and c-2, no Mg-St particle is found on the surface. Namely, the ordered mixtures can be regarded to be encapsulated by Mg-St. Figure 7 shows the Mg-St-coated ordered mixtures and their surface appearances after adding 0.3 g Mg-St to the ordered mixtures and mixing for 5, 30, and 60 min at 70°C, respectively. Only in the 5-min mixing sample in Figure 7a-1 do a few Mg-St particles remain free and not adhere to the ordered mixtures. In the other photos, there are no free Mg-St particles. As in the effect of preparation temperature (see Figure 6), the ordered mixtures can be regarded as being encapsulated after mixing for a long time, i.e., 60 min.

Figure 8 shows the Mg-St-coated ordered mixtures and their surface appearances after adding Mg-St in three different amounts (0.1, 0.3, and 0.5 g) and mixing for 30 min at 50°C. Only the sample with 0.5 g Mg-St in Figure 8c-1 has free Mg-St particles. The sample prepared with 0.1 g Mg-St has the smoothest surface of the three. The increase of the amount of Mg-St added makes the surface rougher because the number of Mg-St particles on the surface which keep their original shape increases.

Similarly, Figure 9 shows the Mg-St-coated ordered mixtures and their surface appearances

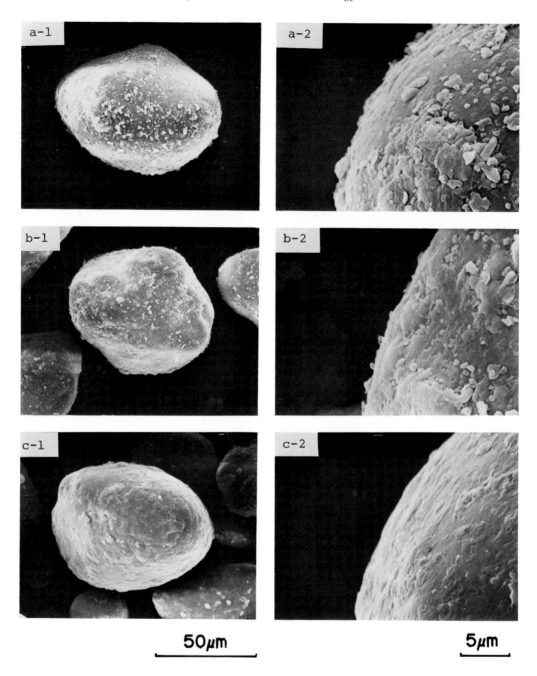

FIGURE 6. Scanning electron micrographs showing changes with temperature of surface appearance of Mg-St-coated ordered mixtures: (a) prepared at 35°C, (b) 50°C, and (c) 70°C. The amount of Mg-St is 0.3 g and mixing time is 60 min.

after mixing with the various amounts of Mg-St for 60 min at 70°C. In none of the cases are free Mg-St particles seen. The surface appearances of Mg-St-coated ordered mixtures are smooth: capsules were made.

Next, the amount of Mg-St adhered to the surface of an ordered mixture, forming a monolayer, is calculated. The particle size and density of the ordered mixtures can be assumed to be almost identical to those of simple potato starch. Assuming that each Mg-St particle

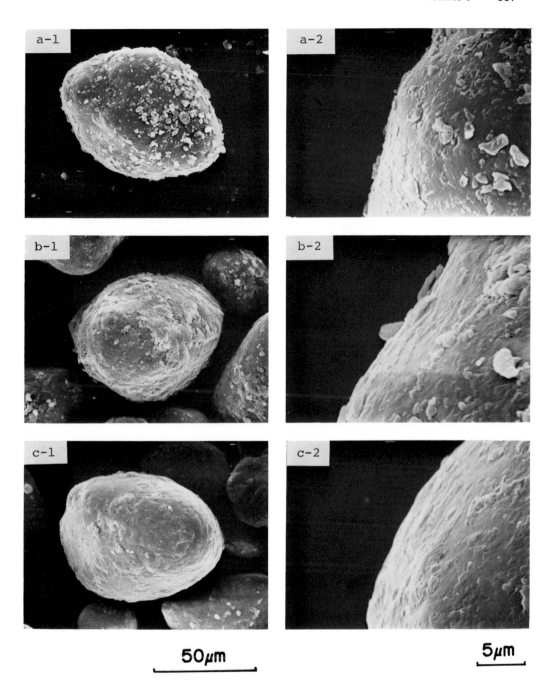

FIGURE 7.   Scanning electron micrographs showing changes with mixing time of surface appearance of Mg-St-coated ordered mixtures: (a) mixed for 5 min, (b) 30 min, and (c) 60 min. The amount of Mg-St is 0.3 g and the temperature is 70°C.

and ordered mixture are spheres and that Mg-St particles are arranged on the ordered mixture, as depicted in Figure 10, the occupied area of one Mg-St particle on the surface of the ordered mixture corresponds to the shaded area, S_1, which is identified by the following equation:

$$S_1 = 2\sqrt{3} \ r_2^2 \qquad (2)$$

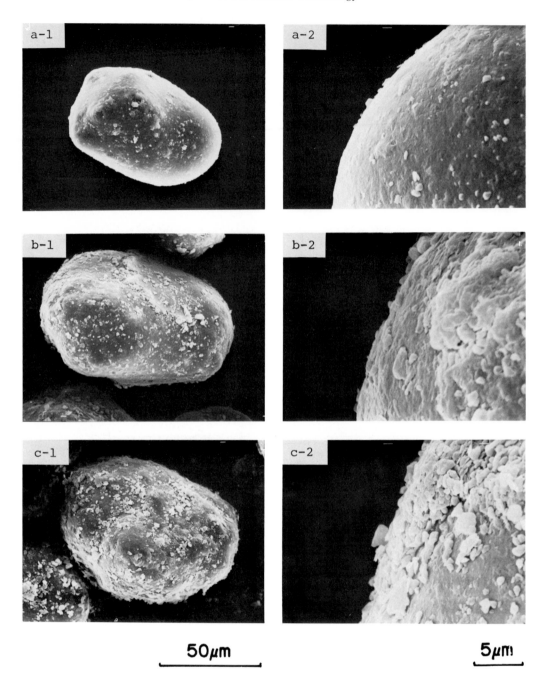

50μm   **5μm**

FIGURE 8.   Scanning electron micrographs showing changes with the amount added of Mg-St of surface appearance of Mg-St-coated ordered mixtures: (a) prepared with 0.1 g Mg-St, (b) 0.3 g Mg-St, and (c) 0.5 g Mg-St. The temperature is 50°C and mixing time is 30 min.

where r_2 is the radius of the Mg-St particle. The specific surface area of potato starch, S_2, is defined by:

$$S_2 = \frac{4\pi r_1^2}{4/3 \times \pi r_1^3 \times d_1} \tag{3}$$

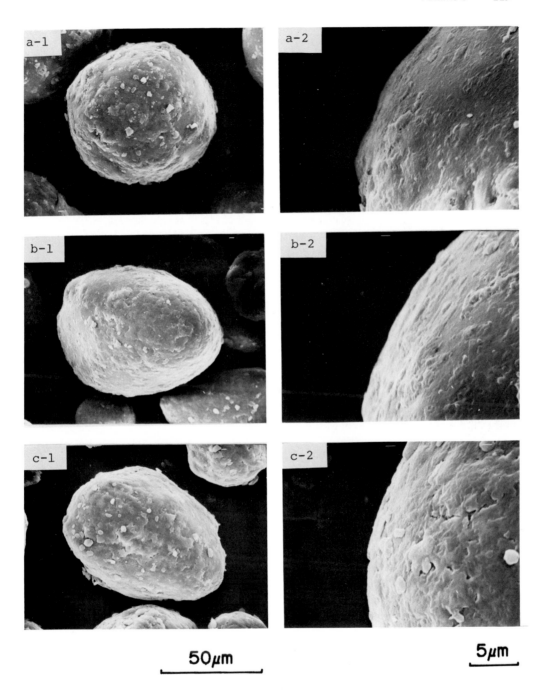

FIGURE 9.   Scanning electron micrographs of encapsulated ordered mixtures and their surface appearances: (a) prepared with 0.1 g Mg-St, (b) 0.3 g Mg-St, and (c) 0.5 g Mg-St. The temperature is 70°C and mixing time is 60 min.

where r_1 and d_1 are the radius and density of potato starch, respectively.[27] Then, the amount of the Mg-St adhered to l g of the ordered mixture, forming a monolayer, is calculated from:

$$\frac{4\pi r_2}{2\sqrt{3}\ r_1} \times \frac{d_2}{d_1} \qquad\qquad (4)$$

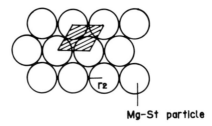

Mg-St particle

FIGURE 10.   Schematic model for arrangement of Mg-St particles on the
surface of potato starch.

where r_1, and r_2, and d_1 are as mentioned above and d_2 is the density of Mg-St. According
to the above equation, the monolayer-coated amount of Mg-St is calculated as 4.3% (w/w).
However, in reality, Mg-St particles would not be arranged as depicted in Figure 10 and a
smaller amount of Mg-St would be sufficient for covering an ordered mixture.

In the case of the sample (prepared with 0.1 g Mg-St), as seen in Figure 8a, a few Mg-
St particles remained free. Hersey et al.[28] called mixtures such as this "partially ordered
randomized mixtures". As mixing time becomes longer or the temperature of the powder
bed becomes higher, the number of free Mg-St particles diminishes and the surface of Mg-
St-coated ordered mixtures becomes smoother. This is caused by heat energy generated
through friction and impact between powders or powder and apparatus, which momentarily
melts the edge of the Mg-St particles or shears them off and causes them to adhere to the
ordered mixtures. Thus, the Mg-St particles on the surface merge with each other to form
a film. While the surface becomes smoother as the supply of energy grows, the demand on
the energy also grows as the number of Mg-St particles increases. Therefore, if the supplied
energy is enough to form a film of Mg-St, the ordered mixtures are encapsulated, independent
of the amount of Mg-St, as seen in Figure 9. However, if an insufficient amount of energy
is supplied, the surface roughness increases as the amount of Mg-St increases.

4. Discussion on Segregation of Ordered Mixtures

Lai and Hersey[29] and Yip and Hersey[31] described that segregation of ordered mixtures
occurred when they were mixed with Mg-St. Yip and Hersey[31] defined two distinct types
of segregation occurring in ordered mixtures: "ordered unit segregation" and "constituent
segregation". Ordered unit segregation occurs in mixtures containing multisized carrier
particles.[32,33] In pharmaceutical mixtures, this leads to drug-rich and drug-lean areas of
powder, even though no change occurs in the distribution of adherent particles on individual
coarse particles. Constituent segregation occurs when fine particles are dissociated from
coarse particles. It occurs when a third component such as Mg-St is added to ordered drug/
carrier mixtures. Lai and Hersey[34] mentioned two possibilities: (1) the third component
adheres preferentially to the carrier particles, displacing the original drug particles from their
adhesion sites, and (2) the third component strips drug particles effectively from carrier
particles, but is not bound itself to the carrier particles. In either case, the homogeneity of
the mixture decreases. However, ordered unit segregation is unlikely to occur in the present
study because potato starch particles, used as carrier particles, have a narrow size distribution
(200 to 250 mesh). Furthermore, the degree of mixing is almost constant and equal to the
value for the ordered mixtures whether Mg-St is present or not (see Figures 4 and 5), and
there is no free particle after adequate mixing (see Figures 6 to 9). Therefore, two models
for the coating of an ordered mixture with Mg-St are possible, as shown in Figure 11b. In
model 11b-1, Mg-St strips off fine isoproterenol HCl particles from the carrier potato starch
particles during the mixing. The agglomerates of Mg-St and isoproterenol HCl adhere again
to the carrier potato starch particles. In this case, some isoproterenol HCl particles exist in

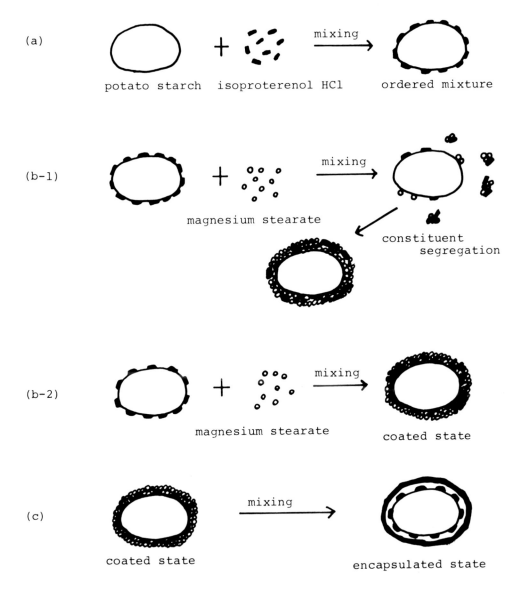

FIGURE 11.   Encapsulation steps of an ordered mixture by Mg-St.

the outer surface layer of the particles. In model 11b-2, Mg-St covers the ordered mixtures, so all isoproterenol HCl particles are assumed to be evenly dispersed over the surface of potato starch particles, just under the Mg-St layer.

In model 11b-1, it is doubtful if the CV value is kept constant after adding Mg-St because model 11b-1 is obtained based on Hersey's first segregation possibility mentioned above.

Moreover, SEM observations of the ordered mixtures show that fine isoproterenol HCl particles adhere to the potato starch surface without keeping their original shapes (see Figure 3). This is the result of heat energy generated during mixing through friction and impact, which melts the edge of the isoproterenol HCl particles or shears them off and causes them to adhere to the potato starch surface. When the drug is adhered in this way, it may be difficult to strip off the drug particles according to model 11b-1,[35] in which the isoproterenol HCl particles on the outer surface of the particles are released first, followed by those in the inner layer. As isoproterenol HCl is a very soluble drug, this release pattern in model

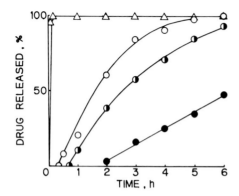

FIGURE 12.   Effect of temperature on drug release after adding Mg-St: ○, prepared at 35°C; ◑, 50°C; and ●, 70°C; △, ordered mixtures. The amount of Mg-St is 0.3 g and mixing time is 60 min. Release tests of the samples were carried out at 37 ± 0.1°C using a paddle method (JP X). 600 mℓ of the first fluid (pH 1.2) was used. The samples (700 mg) filled a gelatin capsule.

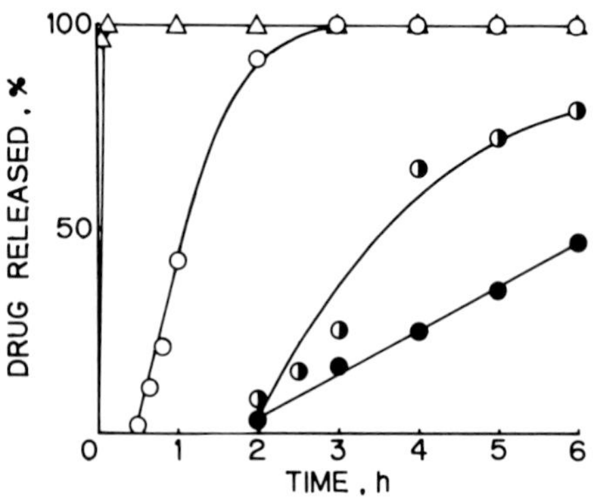

FIGURE 13.   Effect of mixing time on drug release after adding Mg-St: ○, mixed for 5 min; ◑, 30 min; ●, 60 min; △, ordered mixtures. The amount of Mg-St is 0.3 g and the temperature is 70°C. See Figure 12 for release test conditions.

11b-1 indicates the "burst effect". On the other hand, in model 11b-2 where the ordered mixtures are coated with Mg-St, all the isoproterenol HCl particles exist under the layer of Mg-St. Then, this release pattern in model 11b-2 indicates the existence of a "lag time" for the drug to diffuse into the layer of Mg-St.

Figures 12 to 15 show drug release from the Mg-St-coated ordered mixtures prepared under various conditions. According to the results of release tests, all the samples exhibit a lag time but not the burst effect. Thus, the adherence mechanism is the one given in Figure 11b-2.

The above three reasons support model 11b-2 in which Mg-St does not segregate the ordered mixtures, but also covers the surface of the ordered mixtures. This may be the third possible way in which a third component is included in ordered mixtures.

5. Discussion on the Change of Isoproterenol HCl by Mechanochemical Operation

In experiments, isoproterenol HCl was fixed on the potato starch surface by mechanochemical operation. Mechanochemical operation often causes physical or chemical change of materials.

If isoproterenol HCl is oxidized or decomposed (i.e., if isopropyl noradrenochrom is produced), the UV spectrum will not exhibit a maximum at 278 nm. In ordered mixtures, no change was seen in the spectrum, suggesting that no chemical change occurred for ordered mixtures. In order to detemine if mechanochemical operation fractured isoproterenol HCl crystals, an attempt was made to measure its X-ray diffraction pattern. However, it was not possible to do so because the isoproterenol HCl content in ordered mixtures or physical mixtures had to be fixed as low as 1% (w/w), which is less than the least measurable content in X-ray analysis.

D. Release Rate of Isoproterenol HCl

Figure 12 shows the effect of temperature of the powder bed during mixing on the release of isoproterenol HCl. Table 1 gives the initial release rates and lag times for various samples. Release rate decreases and lag time increases with rising preparation temperature. These results indicate that the encapsulated sample prepared at 70°C which has the smoothest surface exhibits the most effective sustained release.

Figure 13 shows the effect of the mixing time on the release of isoproterenol HCl from

Table 1
EFFECT OF TEMPERATURE ON DRUG RELEASE

Temperature (°C)	$k \times 10^3$ (/min)	Correlation coefficient	t_{50} (min)	Initial release rate (%/min)	Lag time (min)
35	11.5	0.996	92	0.42	31
50	7.11	0.996	148	0.39	51
70	2.77	0.995	371	0.17	121

Note: The amount of Mg-St is 0.3 g and mixing time is 60 min.

Table 2
EFFECT OF MIXING TIME ON DRUG RELEASE

Mixing time (min)	$k \times 10^3$ (/min)	Correlation coefficient	t_{50} (min)	Initial release rate (%/min)	Lag time (min)
5	31.2	0.996	62	0.93	40
30	6.80	0.980	219	0.49	117
60	2.77	0.995	371	0.17	121

Note: The amount of Mg-St is 0.3 g and the temperature is 70°C.

the Mg-St-coated ordered mixtures. Table 2 gives the initial release rates and lag times obtained in this study. Similar results were obtained when the effect of preparation temperature was measured. Figure 14 shows the effect of the amount of Mg-St used on the release of isoproterenol HCl. These samples were prepared using a mixing time of 30 min at 50°C. Table 3 gives the initial release rates and lag times for this experiment. Some previous works[19,23,24] report that with an increase in the amount of Mg-St, the release rate decreases. In this experiment, however, the encapsulated sample with the least amount of Mg-St has the slowest initial release rate. EM observations of samples A (prepared with 0.1 g Mg-St at 50°C with 30-min mixing) and B (0.5 g Mg-St at 50°C with 30 min mixing) after release tests for 10 hr show that in sample A a Mg-St film covers almost all the surface of the potato starch, whereas in sample B some flake-like masses of Mg-St are seen on the surface (see Figure 16).

Figure 15 also shows the effect of the amount of Mg-St added on the release of isoproterenol HCl. These samples were prepared using a mixing time of 60 min at 70°C. For all samples encapsulation is achieved independent of the amount of Mg-St. Table 4 gives the initial rates and lag times. The sample with the highest amount of Mg-St has the longest lag time and the slowest initial release rate. EM observations of samples C (prepared with 0.1 g Mg-St at 70°C with 60-min mixing) and D (0.5 g Mg-St at 70°C with 60 min mixing) after release tests reveal that neither sample is destroyed and both almost keep the shell shape (see Figure 16).

Consequently, the release of isoproterenol HCl from Mg-St-coated ordered mixtures is likely to be controlled by the Mg-St film. Moreover, this Mg-St layer, as a film, has fine cracks or small capillary-like pores. The possible mechanism of drug release from Mg-St-coated ordered mixtures, shown in Figure 17a, would resemble that of the system where a partially soluble membrane encloses a drug core, shown in Figure 17b.

Dissolution of part of the membrane allows for diffusion of the constrained drug through the pores in the polymer coat.

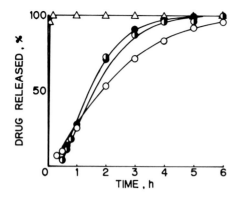

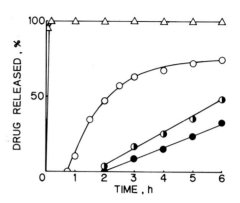

FIGURE 14. Effect of the amount added of Mg-St on drug release: ○, prepared with 0.1 g Mg-St; ◑, 0.3 g Mg-St; and ●, 0.5 g Mg-St; △, ordered mixtures. Temperature is 50°C and mixing time is 30 min. See Figure 12 for release test conditions.

FIGURE 15. Effect of the amount added of Mg-St on drug release: ○, prepared with 0.1 g Mg-St; ◑, 0.3 g Mg-St; and ●, 0.5 g Mg-St; △, ordered mixtures. Temperature is 70°C and mixing time is 60 min. See Figure 12 for release test conditions.

Table 3
EFFECT OF AMOUNT OF Mg-St ON DRUG RELEASE

Mg-St (g)	k × 10³ (/min)	Correlation coefficient	t₅₀ (min)	Initial release rate(%/min)	Lag time (min)
0.1	7.86	0.998	107	0.33	19
0.3	14.6	0.998	81	0.69	34
0.5	15.5	0.998	79	0.73	34

Note: The temperature is 50°C and the mixing time is 30 min.

Table 4
EFFECT OF AMOUNT OF Mg-St ON DRUG RELEASE AND THICKNESS OF Mg-St FILM

Mg-St (g)	k × 10³ (/min)	Correlation coefficient	t₅₀ (min)	Initial release rate (%/min)	Lag time (min)	Thickness of Mg-St film (μm)
0.1	7.56	0.994	133	0.77	41	0.27
0.3	2.77	0.995	371	0.17	121	0.81
0.5	1.97	0.990	501	0.13	145	1.36

Note: The temperature is 70°C and the mixing time is 60 min.

In this case, the data are expressed by the following equation:

$$-\frac{dC}{dt} = \frac{AD}{l}(C - C_m)$$ (5)

where $\frac{dC}{dt}$ is the release rate, A is the area, D is the diffusion coefficient, and l is the thickness of the membrane. C is the drug concentration in the core and C_m is that in the surrounding

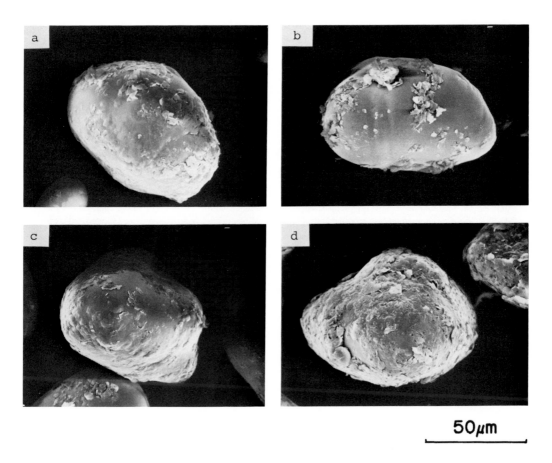

FIGURE 16. Scanning electron micrographs of Mg-St-coated ordered mixtures after release test for 10 hr; a, b, c, and d are, respectively, samples A, B, C, and D.

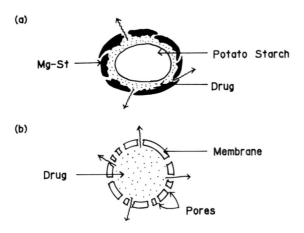

FIGURE 17. Diffusion control of drug release: (a) a model of encapsulated ordered mixture and (b) a model having a partially water-soluble polymer membrane.

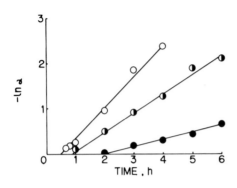

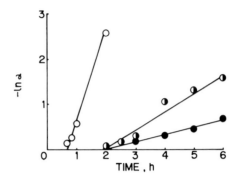

FIGURE 18.   First-order plots of drug release from Mg-St-coated ordered mixtures: ○, prepared at 35°C; ◑, 50°C; and ●, 70°C. The amount of Mg-St is 0.3 g and mixing time is 60 min.

FIGURE 19.   First-order plots of drug release from Mg-St-coated ordered mixtures: ○, prepared using mixing time of 5 min; ◑, 30 min; and ●, 60 min. The amount of Mg-St is 0.3 g and the temperature is 70°C.

medium. Assuming $\dfrac{AD}{l} = k = $ constant and $C \gg C_m$, Equation 5 can be rewritten as:

$$-\frac{dC}{dt} = kC \tag{6}$$

and then Equation 6 is integrated to give:

$$\ln C = \ln C_0 - kt \tag{7}$$

where C_0 is the initial concentration and t is time. On the other hand, residual ratio, α, is defined by:

$$\alpha = \frac{C}{C_0} \tag{8}$$

Then, Equation 7 can be rewritten as:

$$-\ln \alpha = kt \tag{9}$$

where k is the first-order rate constant. Therefore, the diffusion controlled release of this type follows first-order kinetics.

   The plots of $-\ln \alpha$ vs. t for all the release tests are shown in Figures 18 to 21. Straight lines are obtained from the start until the drugs were more than 90% released. First-order rate constants, k, correlation coefficients, and $t_{50}s$ (time needed for 50% drug release) are listed in Tables 1 to 4, respectively. It is seen from these results that all the rates follow first-order kinetics. Therefore, it can be said that the release of the drug from Mg-St-coated ordered mixtures, illustrated in Fig. 17a, is diffusion controlled.

   In model b, the fraction of soluble polymer in the membrane is the dominant factor in controlling the drug release rate. Similarly, in model a, the appearance of the particle is the dominant factor. Namely, the situation in which many Mg-St particles with their original shapes exist on the surface is the same as that in which the fraction of soluble polymer in the membrane is large and there are many pores in the membrane.

   Furthermore, if AD is assumed constant, then k is inversely proportional to *l*. For samples C, E (prepared with 0.3 g Mg-St at 70°C with 60-min mixing), and D which are encapsulated

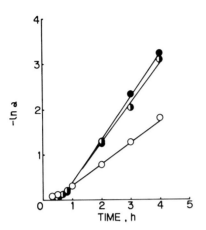

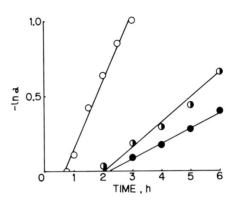

FIGURE 20. First-order plots of drug release from Mg-St-coated ordered mixtures: ○, prepared with 0.1 g Mg-St; ◑, 0.3 g Mg-St; and ●, 0.5 g Mg-St. Temperature is 50°C and mixing time is 30 min.

FIGURE 21. First-order plots of drug release from Mg-St-coated ordered mixtures: ○, prepared with 0.1 g Mg-St; ◑, 0.3 g Mg-St; and ●, 0.5 g Mg-St. Temperature is 70°C and mixing time is 60 min.

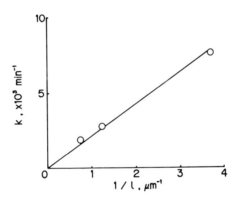

FIGURE 22. Relationship between the first-order rate constant, k, and the reciprocal thickness of Mg-St film, $1/l$.

sufficiently, the thickness of the Mg-St film is calculated by the following equation, on the assumption that ordered mixtures and potato starch have the same radius and the membrane is dense:

$$l = \frac{4/3 \times \pi r_1^3 \times d_1}{4\pi r_1^2} \frac{f}{d_2} \tag{10}$$

where r_1, d_1, and d_2 are as mentioned above and f is the fraction of Mg-St in Mg-St-coated ordered mixtures. The values of l are listed in Table 4. Figure 22 gives the plots of first-order rate constant k vs. $1/l$. Figure 22 illustrates that k is inversely proportional to l. Consequently, this result also indicates that the release of drug from the encapsulated ordered mixtures is diffusion controlled.

E. Conclusion

Ordered mixtures can be prepared by mixing fine cohesive isoproterenol HCl particles and coarser potato starch. If the frictional pressure during mixing was strong enough to shear off isoproterenol HCl, the drug particles were fixed on the surface of potato starch. Then, Mg-St was added and mixed with the ordered mixtures, but isoproterenol HCl was not stripped off from the surface of the potato starch; constituent segregation of ordered mixture did not happen. On the contrary, Mg-St covered the ordered mixtures.

As the supplied mechanochemical energy increased, the Mg-St layer became denser and the surface of the layer was made smoother. Mg-St-coated ordered mixtures showed a sustained release effect. This effect depended on the density of the Mg-St layer but not on the amount of Mg-St alone. Only in the case of encapsulated ordered mixtures did the drug release rate diminish as the amount of Mg-St added increased. Furthermore, all the release rates followed first-order kinetics. The first-order rate constant for encapsulated ordered mixtures was inversely proportional to the thickness of the Mg-St film. Namely, Mg-St-coated ordered mixtures can be considered a diffusion-controlled system with a film having some pores. Finally, this type of product is safe because no organic solvent is used in its preparations. Accordingly, this preparation method is expected to find new pharmaceutical applications in the future.

III. MECHANOCHEMICAL ENCAPSULATION OF A GRANULE AND ITS APPLICATIONS TO SUSTAINED RELEASE PREPARATIONS

A. Fundamental Considerations of Granule Surface Modification

The control of drug release rate from various dosage forms is an important consideration in the drug delivery of ingested medications. The drug release rate depends on the disintegration rates for various dosage forms and the dissolution rate of the drug involved. Therefore, accurate regulation of the desired rates is necessary for controlling drug release rate.

Studies have been made to control the drug release rate by drug surface modifications. Lerk et al.[36] attempted to make the surface of hydrophobic hexobarbital particles hydrophilic by coating the particles with such binders as methylcellulose and hydroxyethylcellulose. They found that the release rate for the drug was appropriately increased. Matsuda et al.[37] also reported that the hydrophobic phenylbutazone surface was made hydrophilic by the adsorption of a surfactant to increase the drug release rate. On the other hand, Ku and Matsumoto[38] showed that deposition of glyceryl monostearate on the surface of hydrophilic nalidixic acid and aspirin particles decreased the release rates for the drugs. These studies indicated the possibility that surface modification to drug particles is a useful method for controlling the drug release rate.

At present, sustained release preparations of granule dosage forms are usually prepared by filling capsules with granules coated with such hydrophobic materials as yellow beeswax or stearyl monoglyceride at various thicknesses. In many cases, however, spray coating is used to cover the surface of granules with hydrophobic materials dissolved in a suitable organic solvent, which may be poisonous to the human body. Hence, there are many problems with the preparation process and the products. In these circumstances, a new pharmaceutical preparation technique that can solve these problems is desired.

When granules are mixed with a powder in a mixer, particles of the powder are observed to adhere to the granule surfaces. When two ingredient powders are mixed, the mixing state is broadly segregated into two categories, random mixing and ordered mixing.[1] The mixture of granules and powder is similar to the ordered mixture. Since the granule surface is covered with particles of the powder in this case, the property of the granule surface is expected to approach that of the powder surface, according to the area and thickness of the powder layer. Namely, it is considered that granules can be surface modified by dry mixing with powders, as discussed in the powder/powder system in Chapter 2. If the property of the granule surface can be regulated and the drug release rate from granules can be controlled

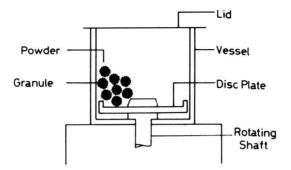

FIGURE 23.   Diagram of centrifugal rotating mixer for mixing of granules and powders. (From Nakagawa, A., et al., *Zairyo Gijutsu*, 2, 43, 1984. With permission.)

by surface modification or encapsulation of the granules by dry mixing with powders, the problem involved in spray coatings now being used can be solved. Dry mixing may thus be a useful pharmaceutical preparation technique.

B. Surface Modification and Encapsulation of Granules
1. Preparation of Granules and Encapsulation of Granules

Lactose granules were prepared from a mixture of α-lactose Japanese Pharmacopoeia (JP) and 4% (w/v) hydroxypropylcellulose ethanol solution by tumbling granulation. The granule products contained 3% (w/w) hydroxypropylcellulose as a binder. The lactose granules for dry blending experiments were obtained by sieving the products through a 16 to 20 mesh. Granules containing acetylsalicylic acid (aspirin) were prepared from a mixture of α-lactose, micronized aspirin, and 4% (w/v) hydroxypropylcellulose ethanol solution by tumbling granulation. The granule products contained 20% (w/w) aspirin and 3% (w/w) hydroxypropylcellulose. The drug-loaded granules for dry blending experiments were obtained by sieving the products through 20 to 32 mesh. Carnauba wax (15-μm diameter powder) and hydroxypropyl methylcellulose (HPMC, 40-μm diameter powder) were used for surface modification and encapsulation of the plain and drug-loaded granules.

Surface modification and encapsulation of granules was carried out by dry blending and granules and a fine powder in a centrifugal rotating mixer (see Figure 23). In the case of lactose granules, 70 g of granules and a powder, weighed at various weight ratios, were homogeneously mixed in a beaker with a spatula and were then introduced into the mixer. The disk plate was rotated at 500 rpm for 60 min. The operation was carried out at 23°C so as to observe the adhesiveness of carnauba wax on the granule surfaces. In the case of lactose granules containing aspirin, each sample was prepared as follows: 70 g of the granules and carnauba wax were homogeneously mixed in a beaker and then blended in the mixer at 500 rpm for 60 min (Sample A). Sample A was transferred into a beaker. After the same amount of carnauba wax as was used for Sample A was added over it, the mixture was blended again in the mixer for 60 min (Sample B). Sample B was then transferred into a beaker. Furthermore, after the same amount of carnauba wax as was used for Sample A was added over it, the sample was blended again in the mixer for another 60 min (Sample C). The same procedures as used in the preparation of Sample C were performed using the mixture of HPMC and carnauba wax, previously blended in a V-type mixer. The HPMC/carnauba wax ratios were kept at 1:20, 2:20, and 4:20 (Samples D, E, and F). The dry mixing for preparing Samples A, B, C, D, E, and F was conducted at 65°C, to obtain a tighter and more smoothly increased adhesion of carnauba wax particles fixed on the granule surfaces.

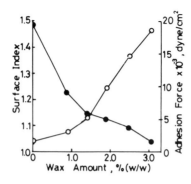

FIGURE 24.   Plots of surface index and adhesion force vs. amount of carnauba wax adhering to granule surfaces: ●, surface index and ○, adhesion force. (From Nakagawa, A., et al., *Zairyo Gijutsu*, 2, 43, 1984. With permission.)

Table 5
ADHESION FORCE OF α-LACTOSE POWDER AND CARNAUBA WAX

Materials	Powder bed porosity	Tensile force (g)	Adhesion force (dyn/cm²)
α-Lactose	0.540	6.58	6.45×10^2
Carnauba wax	0.537	49.42	48.43×10^2

From Nakagawa, A., et al., *Zairyo Gijutsu*, 2, 43, 1984. With permission.

2. Discussion on Granule Surface Modification and Encapsulation

A lactose granule was used as a model granule. When it came into contact with water, it immediately disintegrated and dissolved. If a drug is dispersed in the granule, it is believed that it, too, would dissolve rapidly. Because hydrophobization of the hydrophilic granule surface was necessary to control the drug release from the granules, carnauba wax was used as a powder for surface modification and encapsulation of the granules. The carnauba wax powder has a contact angle of 98.4° for water. Figure 24 shows the surface index of a lactose granule and the adhesion force among lactose granules as a function of the amount of carnauba wax adhering onto granule surfaces. The surface index gradually decreased with the increasing carnauba wax amount. This finding indicates that the shape of the granule gradually changed and approached a complete sphere.

On the other hand, the adhesion force increased with an increase in the carnauba wax amount. Table 5 shows that carnauba wax is more adhesive than lactose powder. The results in Figure 24 and Table 5 reflect the increased granule surface area covered with carnauba wax. Furthermore, the surface appearances of the granules treated with carnauba wax were observed by means of SEM (Figure 25). Many rough gaps or cracks among lactose crystals were seen on the bare granule surface (Figure 25a). Lactose granules treated with 0.9% (w/w) or 1.4% (w/w) carnauba wax (Figure 25b and c) have gaps or cracks that are filled with fine carnauba wax particles. The granule surfaces treated with 1.9, 2.5, and 3.1% (w/w) carnauba wax were partially or completely covered with carnauba wax, (see Figure 25d, e, and f). The modified surfaces are thus rather smooth, compared with the bare granule surfaces. The results for surface modification in Figure 25 indicate that carnauba wax multilayers are formed by many fine carnauba wax particles, after formation of an ordered

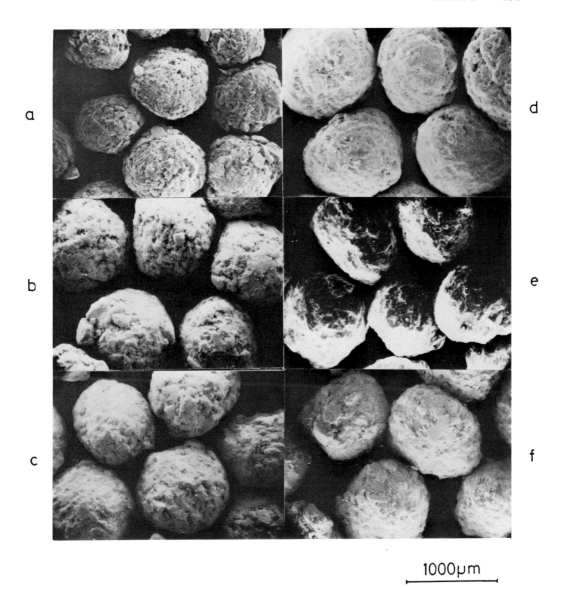

1000μm

FIGURE 25.   Scanning electron micrographs of bare and carnauba wax-modified lactose granules: (a) bare granule, (b) granule modified with 0.9% (w/w) carnauba wax, (c) 1.4% (w/w), (d) 1.9% (w/w), (e) 2.5% (w/w), and (f) 3.1% (w/w). (From Nakagawa, A., et al., *Zairyo Gijutsu*, 2, 43, 1984. With permission.)

mixture, over the granule surfaces. This is supported by the data on changes in the surface index and adhesion force, shown in Figure 24.

In a more detailed discussion, encapsulation of the granule was observed after dissolution of the core materials, as shown in Figure 26. From these photos, it is obvious that continuous carnauba wax shells are formed over the granule surfaces at 2.5% (w/w) carnauba wax. Judging from other photos (not shown here), the carnauba wax shell thickness ranged from 2.4 to 10.5) μm at 2.5% (w/w) carnauba wax.

These experimental results suggest that carnauba wax multilayers are partially formed on the granule surface at a small amount of carnauba wax (up to 1.9% [w/w]). It is interesting to note that the granules are clearly encapsulated at 2.5% (w/w) carnauba wax. Illustrating these results, Figure 27 shows typical SEM photographs of a 10% (w/w) carnauba wax-

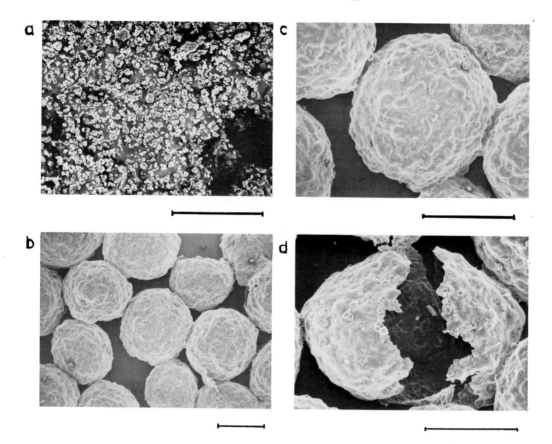

FIGURE 26.   Scanning electron micrographs of carnauba wax particles and carnauba wax capsules: (a) carnauba wax, (b and c) surface appearances of capsules, and (d) surface and inside appearances of capsules. Scales = 500 μm. (From Nakagawa, A., et al., *Zairyo Gijutsu*, 2, 43, 1984. With permission.)

encapsulated microcapsule containing 4% (w/w) isoproterenol HCl-coated potato starch particle, shown in the same figure. The coating on the particle appears to be thick and uniform, confirming the belief that encapsulation of particles takes place by mutual aggregation and their melting of many wax particles around the core particle.

Next, the carnauba wax layer formation mechanism was investigated. Changes in the surface index of lactose granules treated with 2.5% (w/w) carnauba wax and the adhesion force among lactose granules with mixing time lapse were examined. The results are shown in Figure 28. The surface index decreased until mixing time reached 40 min and then leveled off. This indicates that the granule shape changes during the first 40 min of mixing and remains unchanged afterwards.

On the other hand, the adhesion force among granules increased with mixing time. The tendency still continued after 40 min mixing, when the surface index became constant. The increase in adhesion force seems to suggest that the carnauba wax multilayers become dense and cover the granule surface homogeneously. To confirm this, the surface appearances of the granules were observed by SEM (Figure 29). Figure 29a shows the ordered mixture of lactose granules and carnauba wax particles at the initial stage. Many fine wax particles are seen on the granule surface. The small gaps or cracks on the bar granule surface were filled by wax particles. Figure 29b, c, d, e, and f indicate that the surface appearance of the granules becomes increasingly smoother as mixing time becomes longer. After 30 min in Figure 29f, original carnauba wax particles were not found on the granule surface and the

P.Starch / Isoproterenol HCl

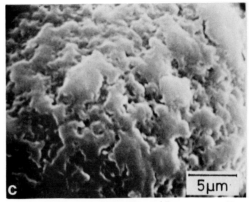

P.Starch / IP HCl / Carnauba Wax

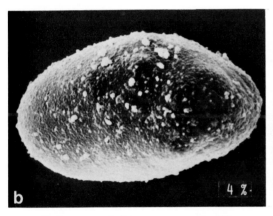

P.Starch / Isoproterenol HCl

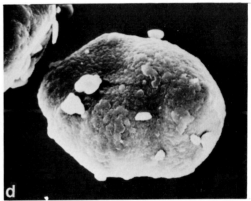

P.Starch / IP HCl / Carnauba Wax

FIGURE 27. Scanning electron micrographs of isoproterenol HCl coated potato starch (a and b) and carnauba wax microcapsule (c and d).

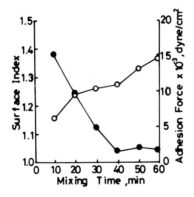

FIGURE 28. Plots of surface index and adhesion force vs. mixing time: ●, surface index and ○, adhesion force. (From Nakagawa, A., et al., *Zairyo Gijutsu*, 2, 43, 1984. With permission.)

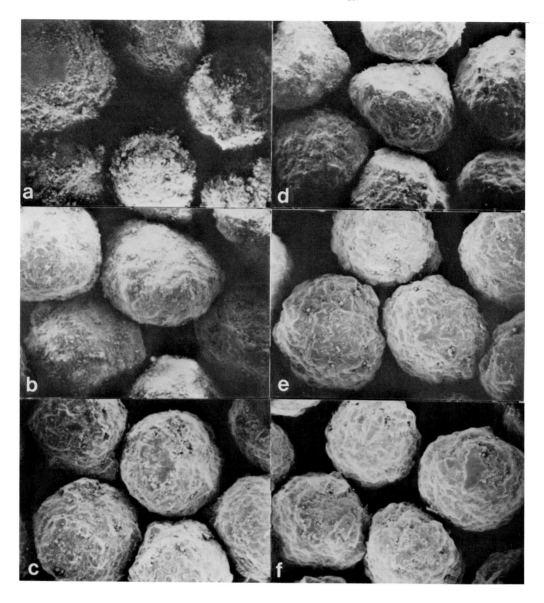

1000µm

FIGURE 29.   Scanning electron micrographs showing change with time lapse in surface appearance of carnauba wax-modified lactose granules: (a) 1 min, (b) 10 min, (c) 20 min, (d) 30 min, (e) 40 min, and (f) 50 min. (From Nakagawa, A., et al., *Zairyo Gijutsu*, 2, 43, 1984. With permission.)

surface appearance was as smooth as that of the capsule in Figure 26a and b. The granules can be regarded as being encapsulated after 50 min mixing.

In the beginning of the mixing stage, fine carnauba wax particles adhere to the granule surface. During the mixing process in the next stage, fine carnauba wax particles with a 75°C melting point were softened by the mechanochemical effects of adherence and/or friction among many wax particles and from friction between wax particles and the disk and/or wall of the mixer. In this way, the granule surface was covered by a soft wax layer, causing

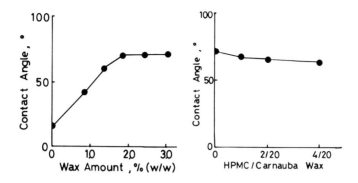

FIGURE 30.   Plots showing extrapolated contact angle of water on tablets prepared from carnauba wax-modified granules: (a) effect of amount of carnauba
wax and (b) effect of amount of HPMC. (From Nakagawa, A., et al., *Zairyo
Gijutsu*, 2, 43, 1984. With permission.)

encapsulation. The softening of wax particles during mixing could be confirmed by temperature elevation data.

Several reports confirm the formation of an ordered mixture in the case of the lactose
granule-carnauba wax system. In the preparation of a powder mixture, Ampolsuk et al.[4]
reported that drugs were spread over the surface of lactose by frictional pressure during dry
mixing. Hersey[8] pointed out that the products obtained by frictional pressure were an ordered
mixture and their drug film over lactose was highly homogeneous. Next, a study was made
to determine how the wetting of the granule surface is changed by dry mixing with carnauba
wax. In all cases, from measurements of the changes, due to time lapse, in the advancing
contact angles of water on tablets prepared from modified lactose granules, it was observed
that the advancing contact angle decreased linearly after an initial rapid decrease. The
apparent contact angle was determined by the extrapolation of advancing contact angles to
time zero. Figure 30a shows the relationship between the extrapolated contact angle and the
amount of carnauba wax. The contact angle increased with the increasing wax amount and
became a plateau at wax amounts higher than 1.9% (w/w). The results indicate that the
wettability of the granule surface could be controlled by changing the carnauba wax amount.
Furthermore, an examination was made to determine if the wettability of lactose granules
could be controlled by the modification of their surface with carnauba wax and a water-
soluble polymer. HPMC was used as the polymer. In all cases, the advancing contact angle
slowly decreased at a constant rate after an initial rapid decrease. The apparent contact angle
was determined by extrapolation of the advancing contact angle to time zero. Figure 30b
illustrates the relation between the extrapolated contact angle and the amount of HPMC.
Since the contact angle decreased slightly with the increasing amount of HPMC, the granule
surface was made slightly hydrophilic.

In general, the observed contact angle of a solid surface composed of two ingredients is
represented by the following equation: $\cos\theta_c = A_1\cos\theta_1 + A_2\cos\theta_2$ where θ_c is the observed
contact angle of a solid surface, θ_1 and θ_2 are true contact angles for each of two ingredients,
and A_1 and A_2 are the fractional areas of two ingredient surfaces of contact angles θ_1 and
θ_2, respectively. The equation indicates that the observed contact angle of a solid surface
depends upon the area ratio of the ingredient surfaces. Accordingly, the contact angle is
expected to decrease as the ratio of water-soluble HPMC to carnauba wax over the granule
surface increases.

Thus, it would be concluded that the wettability of the lactose granule surface can be
controlled by changing the amount of HPMC in the wax matrix adhering to the granule
surface.

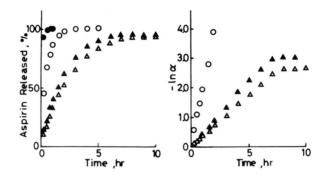

FIGURE 31. Effect of amount of carnauba wax on aspirin release from carnauba wax-modified granule: (a) aspirin released percent vs. time and (b) first-order plots. ●, Nonmodified granule; ○, 2.6% (w/w) carnauba wax-modified granule; ▲, 6.4% (w/w); and △, 10.6% (w/w). (From Nakagawa, A., et al., *Zairyo Gijutsu,* 2, 43, 1984. With permission.)

Table 6
EFFECT OF AMOUNT OF CARNAUBA WAX ON ASPIRIN RELEASE FFROM MODIFIED GRANULES

Sample granules	Wax amount % (w/w)	k (1/hr)	Correlation coefficient	t_{50} (min)
A	2.6	1.88	0.998	19
B	6.4	0.45	0.998	86
C	10.6	0.35	0.999	113

From Nakagawa, A., et al., *Zairyo Gijutsu,* 2, 43, 1984. With permission.

C. Aspirin Release Test

The relationship between the amount of carnauba wax adhering to the granule surface and the aspirin release rate from the drug-loaded lactose granules is shown in Figure 31a. The release rate decreased with the increasing carnauba wax amount. The rate followed the first-order equation, $-\ln \alpha = kt$, where α, k, and t are the residual ratio of drug in the granule, the first-order rate constant, and release time, respectively. In figure 31b plots of $-\ln\alpha$ vs. t are given. Straight lines were obtained from the start until more than 90% of the drug had been released. The data in Table 6 give k, the correlation coefficient, and t_{50} (time of 50% drug release). Figure 32 shows the appearance of the modified granules, before and after the release test. The disintegration rate for the modified granules (2.6% [w/w] carnauba wax) was slower than that for the bare granules. A small amount of carnauba wax capsules was seen to remain in the vessel (Figure 32a-2). Accordingly, the release of aspirin from these granules would be caused by disintegration of granules and subsequent dissolution of the drug. On the other hand, the modified granules (6.4 and 10.6% [w/w] carnauba wax) were found almost encapsulated (Figure 32b-2 and c-2). The release of aspirin from the latter two modified granules would be as in Figure 33, which illustrates an aspirin release model from the encapsulated granules. At first, the release medium penetrates into the granules through channels in the carnauba wax shell. Core lactose then changes its state through wetting into a pasty sludge. Therefore, pasty lactose containing aspirin may be released out of the capsule through channels in the carnauba wax shell. It was reported that

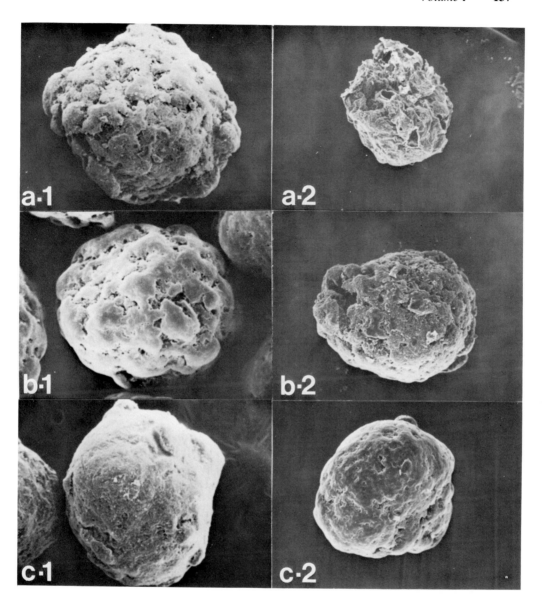

500µm

FIGURE 32.   Scanning electron micrographs of lactose granules containing aspirin modified with 2.6% (w/w) carnauba wax (a-1 and 2), 6.4%(w/w) (b-1 and 2), and 10.6%(w/w) (c -1 and 2) before (1) and after (2) release test. (From Nakagawa, A., et al., *Zairyo Gijutsu*, 2, 43, 1984. With permission.)

the increase in the shell thickness of the microcapsule decreased the drug release.[39] Actually, the aspirin release rate from the modified granules (10.6% [w/w] carnauba wax) was observed to be smaller than that from the modified granules (6.4% [w/w] carnauba wax) due to the thickness effect of the shell. Next, the effect of the amount of HPMC in carnauba wax, previously blended, on aspirin release was studied. The relationship between the amount of HPMC and the aspirin release rate is shown in Figure 34a. The release rate increased with the increase in the amount of HPMC. In this case, as can be seen in Figure 34b, the rate followed first-order kinetics over a 7-hr period from the start. The values of k, the correlation

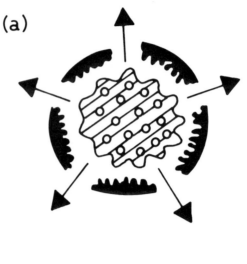

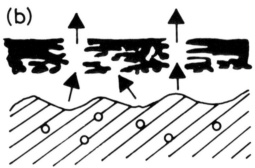

FIGURE 33.   Model of aspirin release from encapsulated granule: (a) general view of capsule and (b) magnified view of capsule. ■, Wax shell and ▨, pasty lactose containing aspirin. (From Nakagawa, A., et al., *Zairyo Gijutsu,* 2, 43, 1984. With permission.)

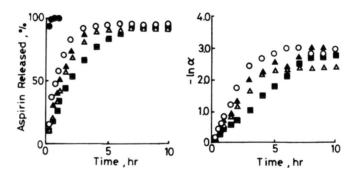

FIGURE 34.   Effect of amount of HPMC on aspirin release from carnauba wax-modified granule: (a) aspirin release percent at different HPMC/carnauba wax ratios vs. time and (b) first-order plots. Weight ratio: ●, 0; ○, 4/20; ▲, 2/20; △, 1/20; and ■, 0/20. (From Nakagawa, A., et al., *Zairyo Gijutsu,* 2, 43, 1984. With permission.)

Table 7
EFFECT OF AMOUNT OF HPMC ON ASPIRIN
RELEASE FROM CARNAUBA WAX-MODIFIED
GRANULES

Sample granules	HPMC/carnauba wax	k (1/hr)	Correlation coefficient	t_{50} (min)
C	0/20	0.35	0.999	113
D	1/20	0.55	0.999	77
E	2/20	0.65	0.997	63
F	4/20	0.83	0.998	49

From Nakagawa, A., et al., *Zairyo Gijutsu,* 2, 43, 1984. With permission.

coefficient, and t_{50} are given in Table 7. Figure 35 shows the surface appearances of the granules (HPMC/carnauba wax weight ratios: 1:20, 2:20, and 4:20) before and after the release test. From these photos, it was seen that all granules were completely encapsulated by the wax matrix. As can be seen in Figure 35c-2, the existence of many pores in the carnauba wax shell is evident after HPMC dissolution. Accordingly, the increase in the HPMC amount in turn increases the number of channels formed by the particles dispersed in the carnauba wax shell, thereby enhancing aspirin release. From the release results, it is clear that the aspirin release rate from the modified granules can be controlled by changing the amounts of carnauba wax and HPMC. The experimental results were summarized as follows: dry mixing of granules and fine powder particles, such as carnauba wax and HPMC, yields ordered mixtures and then converts the granules into a completely encapsulated form. The aspirin release rate from the modified granules can be accurately controlled by changing the amounts of carnauba wax and HPMC.

D. Conclusion

Lactose granules and carnauba wax or HPMC/carnauba wax were blended in a centrifugal rotating mixer. The surface appearances of the modified granules were observed by SEM and the surface modification degree was evaluated by measurements on (1) adhesion force, (2) surface index, and (3) contact angle. Two main factors resulted: (1) smooth and continuous wax multilayers were formed over the granule surface, which was hydrophobic to different degrees with different weight ratios of wax to core granules, and (2) modified granules were made slightly hydrophilic by increasing HPMC in the HPMC/wax combination. Aspirin release from lactose granules, modified and encapsulated with carnauba wax or HPMC/wax, was examined. The release tests using a rotating basket method (JP X) were carried out at 100 rpm in the first fluid (pH 1.2, JP X) at 37°C. The aspirin release rate from lactose granules modified with wax decreased with the increasing wax amount. On the contrary, the release rate from lactose granules, modified with HPMC/wax, increased slightly with the HPMC increase in the HPMC/wax combination. In all cases studied, the aspirin release rate followed first-order kinetics.

ACKNOWLEDGMENT

The authors thank Miss Hisae Yoshizawa for her technical help and research experiments, compiled in her master's thesis entitled Mg-St-Coated Ordered Mixtures.

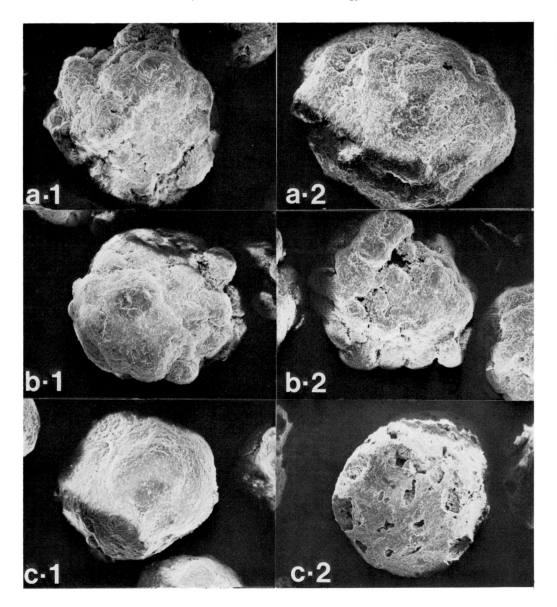

$\overline{}$ 500μm

FIGURE 35.   Scanning electron micrographs of lactose granules containing aspirin modified with 1/20 mixture of HPMC and carnauba wax in weight ratio (a-1 and 2), 2/20 (b-1 and 2), and 4/20 (c-1 and 2), before (1) and after (2) release test. (From Nakagawa, A., et al., *Zairyo Gijutsu,* 2, 43, 1984. With permission.)

REFERENCES

1. **Hersey, J. A.,** Ordered mixing: a new concept in powder mixing practice, *Powder Technol.,* 11, 41, 1975.
2. **Thiel, W. J. and Nguyen, L. T.,** Fluidized bed granulation of an ordered powder mixture, *J. Pharm. Pharmacol.,* 34, 692, 1982.
3. **Thiel, W. J. and Nguyen, L. T.,** Fluidized bed film coating of an ordered powder mixture to produce microencapsulated ordered units, *J. Pharm. Pharmacol.,* 36, 145, 1984.
4. **Ampolsuk, C., Mauro, J. V., Nyhuis, A. A., Shah, N., and Jarowski, C. I.,** Influence of dispersion method on disoolution rate of digoxin-lactose and hydrocortisone-lactose triturations. I, *J. Pharm. Sci.,* 63, 117, 1974.
5. **Lacey, P. M. C.,** The mixing of solid particles, *Trans. Inst. Chem. Eng.,* 21, 53, 1943.
6. **Poole, K. R., Taylor, R. F., and Wall, G. P.,** Mixing powders to fine-scale homogeneity: studies of batch mixing, *Trans. Inst. Chem. Eng.,* 42, T305, 1964.
7. **Travers, D. N. and White, R. C.,** The mixing of micronized sodium bicarbonate with sucrose crystals, *J. Pharm. Pharmacol.,* 23, 260S, 1971.
8. **Hersey, J. A.,** Powder mixing by frictional pressure: specific example of use of ordered mixing, *J. Pharm. Sci.,* 63, p. 1960, 1974.
9. **Staniforth, J. N. and Rees, J. E.,** Powder mixing by triboelectrification, *Powder Technol.,* 30, 255, 1981.
10. **Hersey, J. A.,** Determination of interparticulate forces in ordered powder mixes, *J. Pharm. Pharmacol.,* 33, 485, 1981.
11. **Staniforth, J. N. and Rees, J. E.,** Electrostatic charge interactions in ordered powder mixes, *J. Pharm. Pharmacol.,* 34, 69, 1982.
12. **Staniforth, J. N., Rees, J. E., Lai, F. K., and Hersey, J. A.,** Interparticle forces in binary and ternary ordered powder mixes, *J. Pharm. Pharmacol.,* 34, 141, 1982.
13. **Egermann, H. and Orr, N. A.,** Ordered mixtures — interactive mixtures, *Powder Technol.,* 36, 117, 1983.
14. **Yeung, C. C. and Hersey, J. A.,** Ordered powder mixing of coarse and fine particulate systems, *Powder Technol.,* 22, 127, 1979.
15. **Crooks, M. J. and Ho, R.,** Ordered mixing in direct compression of tablets, *Powder Technol.,* 14, 161, 1976.
16. **Johnson, M. C. R.,** Powder mixing in direct compression formulation by ordered and random processes, *J. Pharm. Pharmacol.,* 31, 273, 1979.
17. **Thiel, W. J. and Stephenson, P. L.,** The effect of humidity on the production of ordered mixtures, *Powder Technol.,* 25, 115, 1980.
18. **Levy, G. and Gumtow, R. H.,** Effect of certain tablet formulation factors on dissolution rate of the active ingredient, III, *J. Pharm. Sci.,* 52, 1139, 1963.
19. **Samyn, J. C. and Jung, W. Y.,** In vitro dissolution from several experimental capsule formulations, *J. Pharm. Sci.,* 59, 169, 1970.
20. **Motycka, S. and Nairn, J. G.,** Influence of wax coatings on release rate of anions from ion-exchange resin beads, *J. Pharm. Sci.,* 67, 500, 1978.
21. **Bolhuis, G. K., Lerk, C. F., Zijlstra, H. T., and Boer, A. H.,** Film formation by magnesium stearate during mixing and its effect on tabletting, *Pharm. Weekbl.,* 110, 317, 1975.
22. **Lerk, C. F. and Bolhuis, G. K.,** Interaction of lubricants and colloidal silica during mixing with excipients, *Pharm. Acta Helv.,* 52, 39, 1977.
23. **Murthy, K. S. and Samyn, J. C.,** Release of capsule formulations containing lubricants, *J. Pharm. Sci.,* 66, 1215, 1977.
24. **Iranloye, T. A. and Parrott, E. I.,** Effects of compression force, particle size, and lubricants on dissolution rate, *J. Pharm. Sci.,* 67, 535, 1978.
25. **Nakagawa, A., Ishizaka, T., Yano, K., and Koishi, M.,** Granule surface modification by dry mixing and applications to sustained release preparations, *Zairyo Gijutsu,* 2, 43, 1984; Reference papers: **Emori, H., Ishizaka, T., and Koishi, M.,** Effects of acrylic acid polymer and its arrangement on drug release from a wax matrix, *J. Pharm. Sci.,* 73, 910, 1984; **Goto, Y., Ishizaka, T., and Koishi, M.,** Effects of water-soluble polymer on sustained release of drug from wax matrix, *Zairyo Gijutsu,* 2, 43, 1984.
26. **Kozatani, J., Kitaura, T., and Ashida, K.,** The influence of dispersing fine granules on degree of mixing, *Yakuzaigaku,* 29, 53, 1969.
27. **Hayashi, H., Kasano, T., and Suhara, K.,** Effect of light anhydrous silicic acid on the angle of repose, packing property, and dispersibility of potato starch powder, *Yakuzaigaku,* 29, 53, 1969.
28. **Hersey, J. A., Thiel, W. J., and Yeung, C. C.,** Partially ordered randomized powder mixtures, *Powder Technol.,* 24, 251, 1979.

29. **Lai, F. K. and Hersey, J. A.,** The variance-sample size relationship and the effects of magnesium stearate on ordered powder mixtures, *Chem. Eng. Sci.,* 36, 1133, 1981.
30. **Bolhuis, G. K. and Lerk, C. F.,** Ordered mixing with lubricant and glidant in tableting mixtures, *J. Pharm. Pharmacol.,* 33, 790, 1981.
31. **Yip, C. W. and Hersey, J. A.,** Segregation in ordered powder mixtures, *Powder Technol.,* 16, 149, 1977.
32. **Jones, T. M.,** The effect of glidant addition of flowability of bulk particulate solids, *J. Soc. Cosmet. Chem.,* 21, 483, 1970.
33. **Thiel, W. J., Nguyen, L. T., and Stephenson, P. L.,** Fluidized bed granulation of an ordered powder mixture reduces the potential for ordered unit segregation, *Powder Technol.,* 34, 75, 1983.
34. **Lai, F. K. and Hersey, J. A.,** A cautionary note on the use of ordered powder mixtures in pharmaceutical dosage forms, *J. Pharm. Pharmacol.,* 31, 800, 1979.
35. **Koishi, M., Ishizaka, T., and Nakajima, T.,** Preparation and surface properties of encapsulated powder pharmaceuticals, *Appl. Biochem. Biotechnol.,* 10, 259, 1984; **Nakajima, T., Ishizaka, T., and Koishi, M.,** Coating of drugs by powder-powder blending method, *J. Pharm. Dyn.,* 7, s-21, 1984; **Nakagawa, A., Ishizaka, T., Yano, K., and Koishi, M.,** Encapsulation of granules by dry blending method and drug release, *J. Pharm. Dyn.,* 7, s-31, 1984.
36. **Lerk, C. F., Lagas, M., and Fell, J. T.,** Effect of hydrophilization of hydrophobic drugs on release rate from capsules, *J. Pharm. Sci.,* 67, 935, 1978.
37. **Matsuda, Y., Kawaguchi, S., Usuki, K., and Eguchi, N.,** Improvement of surface characteristics of a hydrophobic drug by use of a surfactant: effect of hydrophilization on the dissolution rates of phenylbutazone capsules, *Yakugaku Zasshi,* 102, 76, 1982.
38. **Ku, Y. S. and Matsumoto, M.,** Effects of glyceryl monostearate on the dissolution of nalidixic acid and aspirin through depositing on the surface of the powdered drugs, *Yakuzaigaku,* 40, 103, 1980.
39. **Merkel, H. P. and Speiser, P.,** Preparation and in vitro evaluation of cellulose acetate phthalate coacervate microcapsules, *J. Pharm. Sci.,* 62, 1444, 1973.

Chapter 7

AIR SUSPENSION AND FLUIDIZED-BED METHODS

Robert P. Giannini and Pramod P. Sarpotdar

TABLE OF CONTENTS

I. PRINCIPLES

The study of air-suspension and fluidized-bed methods in the manufacture of controlled release systems is a multifaceted endeavor. It requires an understanding of the behavior of solid particles in a stream of gas and how this is affected by equipment design considerations.[1,2] This involves application of the principles of fluid mechanics and heat transfer. It is a study of granulation, granulation followed by coating, and small-particle and tablet coating. It touches the pharmaceutical, chemical, agricultural, food, and energy industries. The literature in the area is immense, as befits a process of such great potential and frustrating complexity. Two reviews have dealt with the process in some detail.[3,4] Also, a number of applications in the biomedical fields appear in a text edited by Franklin Lim.[5] Yet, when reduced to its simplest form, the concept is deceptively easy to understand. All fluidized-bed processes are subject to the basic principles of and problems surrounding the creation of the fluid bed and its maintenance during liquid addition and subsequent drying. The process begins with solid particles of a pure substance or a mixture of ingredients. The particles may be fine powders, pellets, or tablets. They become buoyed (fluidized) in a column of preheated air. An atomized stream of liquid is admitted into the fluid bed which serves to granulate and/or coat the particles, depending on the particulars of the operation. This liquid may be organic or aqueous. It may or may not contain a granulating or film-forming agent. The flow of the liquid may be from directly above the bed, countercurrent to the drying/fluidizing air flow. The liquid flow may also be from beneath the bed, concurrent with the air flow or tangential to the particle flow as in a rotating disk granulator.

II. PROCESS

The first practical requirement is the production of a fluidized bed of dry particles. This aspect of the process has been outlined in several reviews according to the two-phase theory of fluidization.[6-8] At process start in a conventional fluidization system, gas is to be spread evenly by an appropriately designed distributor plate through a packed bed. Each particle is in contact with its closest neighbors. As the gas velocity is increased, a point is reached where the upward force on the particles equals their apparent weight. The bed begins to expand: closest neighbors no longer contact each other and are buoyed by the gas. The gas velocity at this point is called the minimum or incipient fluidization velocity. As gas velocity is increased, excess gas begins to pass through the bed as bubbles. The bed then consists of a lean phase (bubbles) and a dense phase. Gas velocity in the dense phase remains close to the incipient fluidization velocity. The bed resembles a rapidly boiling liquid. The bubbles coalesce as they rise through the bed. The walls of the container affect the shape and rate of rise of bubbles greater than one half its diameter. Figure 1 shows the relationship between height-to-diameter ratio, gas velocity, and gas flow pattern. Bubbles are responsible for the quality of particle mixing in the fluid bed. They also significantly influence the degree of gas/solid contact. When bubbles rise faster than the incipient fluidization velocity, gas circulates within them and comes in contact with only a small amount of solids. Under these conditions mathematical expressions relating liquid and air flow rates, which assume that outlet air temperature and bed temperature are the same, do not apply.[9] Although the two-phase theory has not yet been universally accepted, it has been applied as a working approximation to problems in fluidization relating to heat transfer, dense-phase flow, and spouted beds.[1]

The Wurster process is a spouted-bed variation on the conventional fluid bed design.[10-16] A schematic diagram of a Wurster coating chamber is shown in Figure 2. Rather than distributing gas evenly across the bed, the velocity of the gas stream is higher at the center of the distributor than at the sides. Solid particles are carried upward by this high-

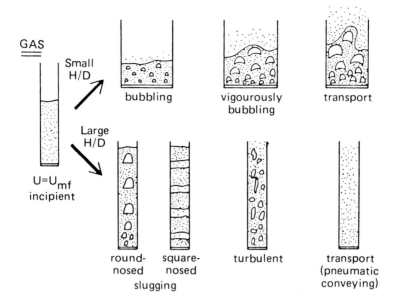

FIGURE 1. Liquid and gas fluidized-bed flow regimes. (From Thiel, W. J., *Int. J. Pharm. Technol. Prod. Manuf.*, 2(4), 6, 1981. With permission.)

velocity stream inside a cylindrical partition. Outside the partition, the gas velocity is lower and particles descend in the annulus formed by the walls of the chamber and the partition. Mixing of particles in this annular region is governed by the same principles of bubble formation and rise as in the conventional fluid bed. Proper circulation of the particles is partly dependent on the distance between the partition and the distributor plate.[17] Normally, the pressure in the region at the mouth of the partition is lower than that in the surrounding annulus near the distributor plate. Thus, solid particles are sucked from the annulus into the mouth of the partition. If the distance between the partition and distributor plate is too large for a given particle size, bulk density and air flow rate, then the pressure differential lessens and sluggish flow results. Fine particles may require a gap on the order of $1/_4$ in., whereas for tablets a gap of 4 in. may be required.

The second practical consideration is the rate at which granulating and/or coating fluids can be applied to a fluid bed. The interactions of air flow, air temperature, drying capacity, affinity of the particles for the solvent system, and chemical stability are subtle and must be carefully studied. During spraying, incomplete mixing in any part of the bed due to a poorly designed distributor or improper gas velocity may result in overwetted bed segments. These segments may defluidize and eventually, if liquid addition is maintained, produce a packed bed through which gas passes only by severe channeling. It is essential during solvent selection to consider data on relative evaporation rate. Some typical data are presented in Table 1, and further information can be obtained from an interesting article by A. A. Sarnotsky.[18]

Solvent-sensitive materials can be granulated or coated by matching the rate of liquid addition and evaporation. When pure water is used, this is easily accomplished with the aid of a psychrometric chart (Figure 3). The addition of a co-solvent, granulating agent, film former, or other additive will reduce the rate of evaporation of the water. The magnitude of this effect is difficult to predict but it can be compensated for by reducing the liquid addition rate or increasing the inlet air temperature. Theoretically a similar procedure can be applied when solvents other than water are used. This is not generally a practical approach since the equivalent of psychrometric charts for a large number of solvents is not readily

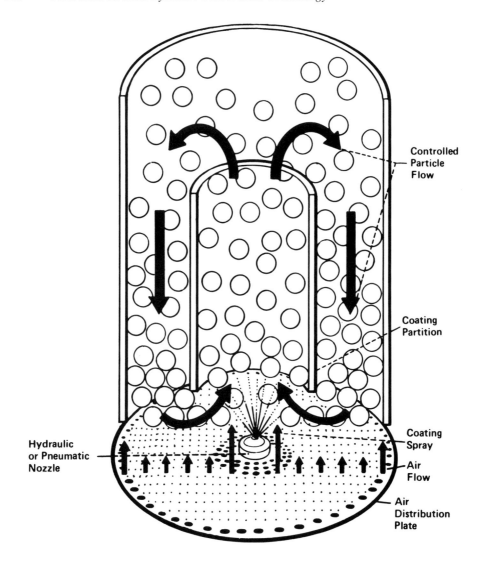

FIGURE 2.   Diagram of a Wurster coating chamber. (From Hall, H. S. and Pondell, R. E., The Wurster process, in *Controlled Release Technologies: Methods, Theory, and Applications*, Vol. 2, Kydonieus, A. F., Ed., CRC Press, Boca Raton, Fla., 1980, 135. With permission.)

Table 1
EVAPORATION RATES[3]
OF SOLVENTS
RELATIVE TO *n*-BUTYL
ALCOHOL = 1.0

Isopropanol, anhydrous	1.7
Ethanol, anhydrous	1.9
Toluene	1.9
Hexane	3.0
Methanol, anhydrous	3.5
Ethyl acetate	4.2
Acetone	7.5

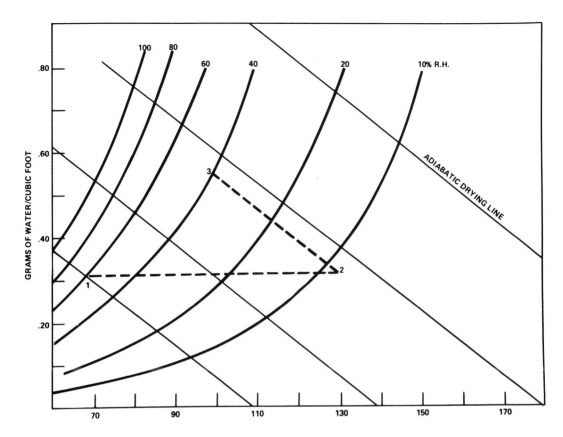

FIGURE 3. Psychrometric chart. (From Hall, H. S. and Pondell, R. E., The Wurster process, in *Controlled Release Technologies: Methods, Theory, and Applications,* Vol. 2, Kydonieus, A. F., Ed., CRC Press, Boca Raton, Fla., 1980, 139. With permission.)

available. However, vapor pressure data are tabulated[19] for many pure solvents over a wide temperature range and it is unlikely that the inlet air supply would contain any organic solvent; therefore an analogous procedure could be applied.

Mixed-solvent systems, however, are the main concern of most pharmaceutical scientists, and the vapor pressure of such mixtures is difficult to predict. Thus, empirical methods for establishing maximum flow rates are widely used. The most efficient of these involves experimental designs in which all variables (i.e., liquid flow rate, inlet air temperature and volume, atomizing air pressure, etc.) are altered at the same time.[20] The techniques of multiple regression[21] and response surface analysis[22] can then be applied to quantitatively evaluate the importance of changes in the individual variables and the level of interaction among them. Frequently such studies reveal that "the fluidizing air temperature and volume, the atomizing air pressure, and the liquid flow rate are the most significant process variables."[23]

A number of valuable observations concerning the fluid-bed process have been assembled in a recent publication.[23] For example:

1. Most fluid-bed granulations are carried out with a top-spray configuration.
2. The bulk density of such granulations is generally lower than that obtained by more traditional techniques.
3. Bulk density increases with increasing batch size. As a rule of thumb, the target for batch bulk density in the laboratory (8 kg or less) should be 20% lower than the desired bulk density of production batches (500 kg or more).

4.   Bulk density can be precisely controlled by concentrating on in-process moisture and fluidizing air volume. The lowest possible bulk density is attained by maintaining the bed moisture content during spraying at or below the maximum finished product moisture.

5.   Granule size has been shown to depend on humidity in the bed during granulation. The effect of ambient air dew-point variation is minimized by high inlet air temperature, but the production inlet air temperature should not exceed those used at laboratory or pilot scale if changes in product performance characteristics due to altered evaporation rates are to be avoided. Moisture analyzers can be used to provide a continuous measure of moisture content during granulation so that process variables can be altered to reproduce a previously established moisture curve.

6.   In the scale-up of fluid-bed granulation processes, the increase in spray rate should be based on the ratio of air flow velocities in the large and small machines rather than on the ratio of batch weights. Manufacturers of production equipment tend to increase the ratio of bed depth to diameter of distribution plate to accommodate larger batches. This allows an adequate fluidization pattern to be achieved without a proportional increase in the volume of fluidizing air. In the absence of integral air speed and volume indicators, the ratio of the cross-sectional areas of the bowl screens can be used to determine the spray rate for the larger production machine.

7.   All the available spraying configurations are used to coat small particles despite the lesser efficiency of the top-spray configuration for coating operations due to a tendency toward spray drying of coating solutions. This is true because available top-spray equipment allows larger batch capacity and is of simpler design than equipment configured for other types of liquid addition. However, its lesser efficiency dictates that the top-spray configuration not be used in situations where physicochemical properties of the finished film are critical determinants of product performance, i.e., controlled release.

8.   Tablet coating is largely confined to Wurster (spouted-bed) coating systems.

9.   Scale-up of spray rate in a coating process must be performed in a more conservative fashion than in granulation. As a result of the high density of particles in the coating zone and the relatively short distance that droplets of coating solution must travel before contacting a particle, there is a tendency to localized agglomeration despite an excess drying capacity in most cases. An additional factor in agglomeration is the general tackiness of most film formers at some point in the drying process. It seems that film formers sprayed from solution are more likely to cause agglomeration than hot melts, latices, or pseudolatices. The easiest way to make use of the excess drying capacity is to add coating zones, i.e., more nozzles in top-spray equipment and more partitions and nozzles in Wurster.

III. MACHINERY

The major elements of concern in the design and construction of a fluid bed system are

1.   Process control systems
2.   Spray systems
3.   Product containers
4.   Charge/discharge systems
5.   Air supply systems
6.   Filters and filter systems
7.   Closed circuits and solvent recovery systems
8.   Explosion prevention, suppression, and control
9.   Cleaning systems

Tables 2 to 5 and Figures 4 to 13 are compilations of equipment specifications, available options, and configurations taken from literature supplied by four major manufacturers of fluid-bed equipment (Aeromatic, Freund, Glatt, and Lakso). All manufacturers offer a similar range of production capacity and machine configurations. All manufacturers have machinery in place and operating satisfactorily in production environments. Some rather obvious differences in dimensions of machinery become apparent as capacity increases: Glatt has greater height-to-diameter ratios; Aeromatic and Vector/Freund have lesser height-to-diameter ratios for coater/granulator equipment. This may affect the performance characteristics of product that needs to be switched from one type of equipment to another, if exactly the same processing conditions are used. However, it seems that in most cases careful adjustments in process variables can overcome the difficulties caused by different fluidization systems due to machine dimensions. One exception might be a product that requires a high fluidization velocity. In such a case the shorter machines may not be suitable. In fact, production-scale equipment is usually built to order and is subject to the limitations of physical plant dimensions. Thus no two pieces of production equipment from the same manufacturer, although of the same nominal capacity, are likely to have exactly equal measurements. This also implies that the dimensions of production equipment from any manufacturer are not directly scaled from their own laboratory or pilot-size equipment. Therefore, "scale-up" is no more or less difficult than switching a product from one make of equipment to another. The same detailed understanding of the critical variables for a particular process and the most likely effect of changing equipment dimensions is required.

IV. APPLICATIONS

Mehta and Jones[24] used a scanning electron microscope to examine the surface of pellets coated in various types of coating equipment. They found that coatings were more uniformly applied by fluid-bed equipment than by either conventional pan or modified perforated pan methods. Among the fluid-bed techniques, the Wurster equipment produced a visibly smoother and more continuous coating than did the top-spray equipment. There was also a distinct difference in the release of a colored marker when pellets produced in the two kinds of equipment were placed in water. Release was slower in Wurster and faster in top-spray, even though the amount of coating material sprayed was the same in both cases and the processing conditions were similar. Coated pellets made using the tangential-spray method seemed to be comparable to those produced by the Wurster column. However, recent data generated in the labs at Himedics indicate that higher bulk densities can be obtained if the core pellet is manufactured in a rotary granulator with tangential spray.

Several commercial applications of fluid-bed technology are known to the authors. In 1982, Key Pharmaceuticals, Inc. launched Theo-Dur Sprinkle®. This product is a controlled release pediatric theophylline micropellet. It is supplied to the market in hard gelatin capsules that are meant to be opened so that the contents can be poured onto soft food. It is manufactured in Wurster columns in a number of steps. The first few steps involve the creation of an active core by successive applications of theophylline to a sugar seed. The final step is the addition of the sustained release coating. The bioavailability of this product is known to be negatively affected by administration with food.[25,26] In 1987, Key Pharmaceuticals, Inc. introduced K-Dur®, a sustained release potassium chloride tablet. The tablet is composed of film-coated crystals of potassium chloride and a minor portion of other pharmaceutical excipients. The in vitro release of potassium is not affected by compression into tablets. This product, manufactured in Wurster columns, is a unique example of formulation science. Materials selection involved matching the viscoelastic properties of potassium chloride with those of the sustained release film coat. Further protection of the coated crystals was provided

Table 2
TECHNICAL SPECIFICATIONS FOR AEROMATIC FLUID BED GRANULATOR/DRYERS

	Size							
	3	4	5	6	7	8	9	10
Loading weight (kg)	15—45	30—90	60—180	100—300	150—450	200—600	300—900	400—1200
Filter area (m²)	2.0	4.5	6.5	11.0	20.0	24.0	36.0	50.0
Base area of product bowl (m²)	0.11	0.25	0.45	0.74	1.11	1.48	2.22	2.93
Capacity of product bowl (m³)	0.118	0.21	0.34	0.54	0.79	1.1	1.48	2.0
Product space bulk (m³)	0.59	1.0	1.5	2.2	3.97	5.2	8.2	11.9
Explosion relief flap area (m²)								
Relief before exhaust air filter	0.15	0.22	0.29	0.37	0.55	0.66	0.89	1.15
Relief after exhaust air filter	0.31	0.38	0.46	1.08	1.46	1.72	1.89	2.46
Relief towards top	0.28	0.5	0.785	1.13	1.4	1.85	2.28	3.42
Amount of air (m³/sec)	0.3	0.6	1.2	2.0	3.0	4.0	6.0	8.0
Static pressure (N/m²)	3000—7000	3000—7000	3000—7000	3000—7000	3000—7000	3000—7000	3000—7000	3000—7000
Motor power (kW) at								
3000 N/m²	2.2	4.0	8.8	11.0	14.7	18.4	29.4	36.7
5000 N/m²	2.9	5.5	11.0	18.4	22.0	29.4	44.1	55.0
7000 N/m²	4.0	7.5	14.7	22	36.8	44.1	73.5	92.0
Heating capacity (kJ/hr) at inlet temperature of fresh air of 263 K (−10°C)								
323 K (50°C)	75,350	155,700	311,400	502,300	753,500	1,004,600	1,507,000	2,009,300
353 K (80°C)	113,000	233,600	467,200	753,500	1,130,200	1,507,000	2,260,400	3,013,900
383 K (110°C)	150,700	311,400	623,700	1,004,600	1,507,000	2,009,300	3,013,900	4,018,600
Maximum spray capacity dm³/min H₂O	2.6	2.6	2.6	2.6	8	8	8	24

From Aeromatic Technical Brochure No. 78.1.M.30, Aeromatic Inc., Towaco, N.J., 1978. With permission.

by judicious selection of the minor pharmaceutical components.

The pharmaceutical division of Pennwalt Corporation has commercialized the concept of coated ion-exchange resin/drug complexes to achieve sustained release oral delivery. A liquid product containing dextromethorphan (30 mg/5 mℓ) is marketed in the U.S. under the trade names of Delsym™ (Pennwalt) and Extend-12™ (A.H. Robins). A combination liquid product containing phenylpropanolamine (equivalent to 37.5 mg/5 mℓ of the hydrochloride salt) and chlorpheniramine (equivalent to 4 mg/5 mℓ of the maleic acid salt) is also available under the trade names of Corsym™ and Cold Factor-12™, both by Pennwalt. Additionally, as of this writing a chlorpheniramine/codeine combination product and a phentermine-containing product are under development. The published technical literature[27-29] concerning the development of this type of product indicates that the final dosage form may be a tablet, capsule, or suspension. However, the major novelty is that stable and palatable sustained release liquid dosage forms can be produced. The products are manufactured in Wurster-type fluid-bed coating equipment using ethyl cellulose (apparently in organic solution) as the film former. As expected, drug release rate both in vitro and in vivo is dependent on particle size, thickness and uniformity of coat, and the nature of the ion-exchange resin core.

Thiel and Nguyen[30] used the technique of ordered mixing to extend the lower size limit for fine particle coating in conventional fluid-bed equipment.

The generally accepted lower limit is about 200 μm, below which agglomeration becomes a serious problem. With specialized equipment Fanger[31] shows that particles as small as 40 μm can be coated. By adsorbing micronized particles (2 to 5 μm) of salicylic acid onto coarser carrier particles (300 to 700 μm) of lactose or dextrose, Thiel and Nguyen made possible the film coating of the ordered units in an Aeromatic (size 1) laboratory fluid-bed unit. The coating solutions used were 5% cellulose acetate phthlate (CAP) in aqueous sodium hydroxide (pH 7 to 8), 3% aqueous hydroxyethyl cellulose, or 8% CAP in a mixed organic-solvent system (50% w/w methylene chloride and methanol). They found that only aqueous coating solutions could be applied to the ordered mixtures directly, retaining between 74 and 95% of the micronized salicylic acid. Organic solutions destroyed the ordered mix entirely. If an aqueous coating of about 2 to 3% by weight was first applied to the ordered mixture, subsequent coating with either aqueous or organic solutions was possible. The stability of the ordered mix during coating was dependent on carrier type, drug load, and the presence or absence of talc. These effects did not seem to be completely consistent with the results of prolonged fluidization studies performed without the addition of any coating solution.

Goodhart et al.[32] coated pellets containing phenylpropanolamine hydrochloride (PPA HCl), microcrystalline cellulose NF, and magnesium stearate NF, using Aquacoat® (a pseudolatex dispersion of ethyl cellulose in water), Eudragit® E-30D (a neutral copolymer based on poly[meth]acrylic esters), or ethyl cellulose from alcoholic solution. The coating was performed in an Aeromatic Model Strea-1. They found that release of PPA HCl from pellets coated with Aquacoat® was more sensitive to cure time and cure temperature when triethylcitrate was used as the plasticizer and less sensitive when dibutyl sebacate (DBS) was used. This was related to the fact that DBS lowered the softening temperature of free films more than did triethylcitrate at equivalent plasticizer levels, indicating that DBS is a more effective plasticizer for Aquacoat® than triethylcitrate. The authors also found that the release of drug from Aquacoat®-coated pellets was pH dependent. Drug release from pellets coated with ethyl cellulose from ethanolic solution was not pH dependent. This difference was attributed to the ionic nature of the stabilizing agent (sodium lauryl sulfate) used in the pseudolatex. It was stated that the release of drug from pellets coated with Eudragit® E-30D was not affected by drying time, drying temperature, or pH.

Senjkovic and Jalsenjak[33] studied the effects of coating isoniazid particles with different

Table 3
TECHNICAL SPECIFICATIONS FOR VECTOR/FREUND FLO-COATERS

	1-bar construction flo-coaters												2-bar construction flo-coaters				
	FL-MINI	FLO-1	FLO-5	FLF-15 FLO-15	FLF-30 FLO-30	FLF-45 FLO-45	FLF-60 FLO-60	FLF-90 FLO-90	FLF-120 FLO-120	FLF-200 FLO-200	FLF-300 FLO-300	FLO-500	FLF-15EX FLO-15EX	FLF-30EX FLO-30EX	FLF-60EX FLO-60EX	FLF-120EX FLO-120EX	FLF-200EX FLO-200EX
Container volume																	
ℓ	0.8	2.3	20	45	100	155	220	300	420	660	1000	1450	45	100	220	420	660
ft^3	0.03	0.08	0.7	1.6	3.5	5.5	7.8	10.6	14.8	23.3	35.3	51.2	1.6	3.5	7.8	14.8	23.3
Working capacity																	
kg	0.4	1.0	5—8	15—22	30—50	45—75	60—100	90—140	120—180	200—280	300—450	550—725	15—22	30—50	60—100	120—180	200—280
lb	0.9	2.2	10—15	35—50	65—110	100—165	130—220	200—300	265—400	440—620	660—1,000	1,210—1,595	35—50	65—110	130—220	265—400	440—620
Container diameter																	
mm	140	190	400	550	720	860	1000	1,100	1,200	1,400	1,600	1,900	550	720	1,000	1,200	1,400
in.	5.5	7.5	16	22	29	34	40	44	48	56	64	76	22	29	40	48	56
Exhaust fan rating — cfm	35	85	280	530	880	1,235	1,765	2,120	2,650	3,530	4,590	5,650	530	880	1,765	2,650	3,530
Exhaust fan rating — kW	0.75	1.9	2.2	3.7	5.5	7.50	11.0	15.0	18.5	22.0	30.0	44.0	3.7	5.5	11.0	18.5	22.0
Heat requirement for 80°C temp. rise Electric Heat Exchanger Btu/hr			46,000	86,400	140,000	201,600	288,000	345,600	432,000	576,000	750,000	922,000	86,400	140,000	288,000	432,000	576,000
Steam consumption per hour																	
Kg			23	42	70	92	141	169	211	282	366	450	42	70	141	211	282
lb			51	92	154	202	310	372	464	620	805	990	92	154	310	464	620
Number of spray guns																	
FLF-peripherally mounted	N/A	N/A	N/A	2	3	3	4	4	4	6	8	N/A	2	3	4	4	6
Flo-center mounted	1	1	1	1	2	2	3	3	3	4	6	9	1	2	3	3	4

Volume I   **153**

Maximum compressed air usage in cfm at 70 psi

FLF	N/A	N/A	24.7	24.7	37	37	50	50	50	74	106	N/A	24.7	37	50	50	74
FLO	4.4	8	17.6	17.6	35.3	35.3	53	53	53	70.6	99	123.5	17.6	35.3	53	53	70.6
Width																	
m	0.5	0.8	1.2	1.31	1.42	1.57	1.71	1.81	1.91	2.21	2.46	2.9	1.31	1.52	1.81	2.01	2.31
ft	1'6"	2'6"	3'9"	4'3"	4'6"	5'1"	5'6"	5'9"	6'2"	7'2"	8'0"	9'5"	4'3"	4'10"	5'9"	6'6"	7'5"
Depth																	
m	0.5	0.6	1.2	1.27	1.53	1.75	1.92	2.05	2.18	2.34	2.77	3.0	1.24	1.53	1.92	2.18	2.38
ft	1'6"	1'9"	3'9"	4'1"	5'0"	5'7"	6'3"	6'7"	7'1"	7'6"	9'0"	9'8"	4'0"	5'0"	6'3"	7'1"	7'8"
Height																	
m	0.6	1.2	1.8	2.34	2.47	2.63	2.78	2.84	2.9	3.31	3.75	4.2	2.34	2.57	2.88	3.23	3.6
ft	1'9"	3'9"	5'9"	7'7"	8'1"	8'6"	9'1"	9'3"	9'5"	10'8"	12'3"	13'7"	7'7"	8'4"	9'4"	10'6"	11'8"
Weight																	
Kg	50	150	600	700	900	1,100	1,300	1,500	1,800	2,200	2,700	6,500	900	1,200	1,700	2,300	2,900
lb	110	330	1,320	1,540	1,980	2,420	2,860	3,300	3,960	4,840	5,940	14,300	1,980	2,640	3,740	5,060	6,380

Note: N/A = not applicable.

From Vector/Freund Technical brochure, Vector/Freund, Marion, Iowa, 1985. With permission.

Table 4

**LAKSO DATA SHEET: TECHNICAL SPECIFICATIONS FOR AIR
SUSPENSION COATING SYSTEMS (WURSTER) — ASCOAT**

Model number	Diameter (in.)	Air volume (cfm)	Static pressure (SP) water (in.)	Capacity (kg/batch)
2	4 × 6	80	18	$^1/_2$—2
4	6	120	20	1—4
25	12	600	24	12—25
40	12 × 18	1,000	30	20—40
80	18	1,500	35	40—80
120	18 × 24	1,800	40	80—120
160	24	3,000	48	120—160
450	46	12,000	50	300—450

Production rates
  Aqueous film   45 min—1 hr
  Solvent film   10 min—30 min
  Sugar coating   15% buildup, 1 hr

Capacity
  Depends greatly upon product

From Lakso Technical brochure, Lakso Company, Leominster, Mass., 1984. With permission.

amounts of ethyl cellulose from ethanolic solution. Using an Aeromatic Model Strea-1, all fluidization parameters were kept constant while increasing volumes of coating solution were applied at a constant rate, with a short drying period between applications. With increasing amounts of coating material the geometric mean diameter of the particles increased and the rate of release of drug decreased. Tablets made from the microcapsules on a hand-operated press and without pharmaceutical additives released drug more slowly than the free micro-capsules. Dissolution data conformed to the square root law.

Lehmann and Dreher[34] have used the fluid-bed technique to coat a number of pharma-ceutical active ingredients in the form of pellets and crystals. Tables 6 and 7 show the range of release-rate characteristics achieved using different Eudragit® acrylic resins. Table 8 shows the results of coating experiments performed on increasing batch sizes of acetylsalicylic acid (ASA) crystals. The coating solution used (based on a 50-kg charge) was Eudragit®L (24.0 kg, 12.5% solution in isopropanol) + dibutylphthalate (0.6 kg) + talc (1.5 kg) + isopro-panol (31.4 kg). The powder bed was fluidized and prewarmed at an inlet air temperature of 55°C for 2 min. The coating solution was then added over a period of 3 hr at an inlet air temperature of 45 to 50°C. The spray nozzle diameter was 1.7 mm. The atomization air pressure was 4 bar. The nozzle was placed approximately 500 mm from the bottom of the dryer. The outlet air temperature was 32°C. At the completion of the spray cycle, heating was discontinued but drying was allowed to proceed for 5 min more.

Similar results were obtained (Table 9) when an aqueous dispersion of Eudragit® L-30D was used without causing a significant decrease in the stability of the ASA. Although a somewhat larger amount of film former was required, as a result of the use of a water-soluble plasticizer, the total spraying time was reduced to 105 min because of the higher solids content of the aqueous dispersion. A high inlet air temperature was also used (54 to 60°C) while the outlet air temperature remained the same (32°C).

ACKNOWLEDGMENTS

The authors gratefully acknowledge the support of Wyeth Laboratories and Key Phar-maceuticals, Inc.

Table 5
TECHNICAL SPECIFICATIONS FOR A GLATT FLUID-BED ROTOR GRANULATOR

Type	Type disc diameter (mm)	Circumferential speed (m/sec) max velocity-min velocity	Disc revolution (n_R/min) number of rpm	Motor capacity (kW) For plane disc	For waffle disc	Air volume (m³/hr)	Differential pressure of turbine (Pa)	Turbine capacity (kW)
GRG 3/5	306	16-3	max 1000 min 200	0.55	0.55	580	4000	1.85
GRG 5/9	480	23-4	max 1000 min 200	1.5	2.2	750	8000	4.0
GRG 15/30	620	25-5	max 735 min 150	4.0	5.5	1500	8000	7.5
GRG 30/60	780	25-5	max 615 min 120	5.5	7.5	3000	8000	15.0
GRG 60/100	1000	25-5	max 500 min 100	7.5	11.0	4500	8000	18.5
GRG 120/200	1400	25-5	max 340 min 70	11.0	15.0	6000	8000	22.0

From Glatt Technical Bulletin #S1283FBRG, Glatt Air Techniques, Inc., Ramsey, N.J., 1983. With permission.

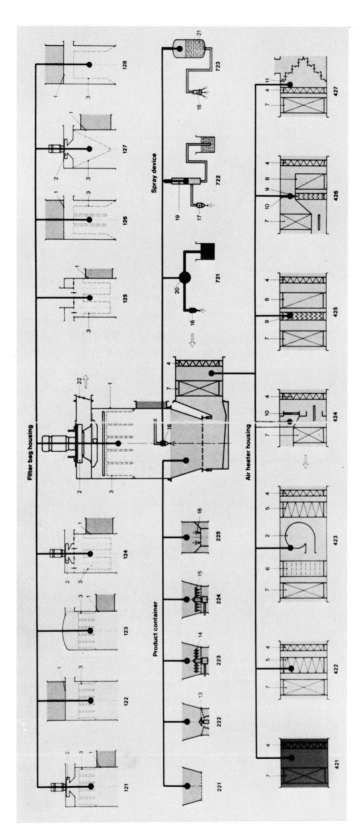

FIGURE 4. Available variations in key components of a fluidized bed system. Key: (1) explosion relief flap, (2) centrifugal fan, (3) exhaust air filter, (4) coarse dust filter G3, (5) fine dust filter F2 or F3, (6) sterile filter S1, (7) air heater, (8) cooler, (9) droplet catcher, (10) bypass flap, (11) blending chamber, (12) heat exchanger, (13) bottom discharge device, (14) finger-shaped agitator, (15) high-speed agitator, (16) two-fluid nozzle, (17) one-fluid nozzle, (18) multihead nozzle, (19) high-pressure pump, (20) dosing pump, (21) liquid tank, (22) air flap, (23) feeding hopper, (24) separator, (25) pressure relief duct, (26) blower unit, (27) protection valve, (28) container tilting device, (29) rapidly acting blender, (30) lifting and swivelling device, (31) tumbler, (32) tablet-compressing machine, (33) discharge container, (34) silo, (35) conveying belt, (36) sluice, (37) weighing unit, (38) sugar mill, (39) distributing switch point, (40) exhaust air duct, (41) inlet air duct, (42) injector, (43) washing nozzle, (44) cleaning liquid, (45) steam duct, (46) water duct, (47) compressed air bottle, (48) ball and nozzle, (49) detector, (50) detonator valve, (51) melt container, (52) product pattern, (53) silencer, (54) pneumatic conveying pipe, (55) conveying worm, (56) condenser, (57) solvent tank, (58) bypass duct, (59) tilting device, (60) extinguisher tank, (61) hot-air feed, (62) discharge sluice, (63) shaking cylinder, (64) air flap to filter blow-off device, (65) pneumatic gasket, (66) product container, (67) heating medium, (68) spray liquid, (69) atomizing air. (From Aeromatic Tech. Brochure No. 78.1.M.30, Aeromatic Inc., Towaco, N.J., 1978. With permission.)

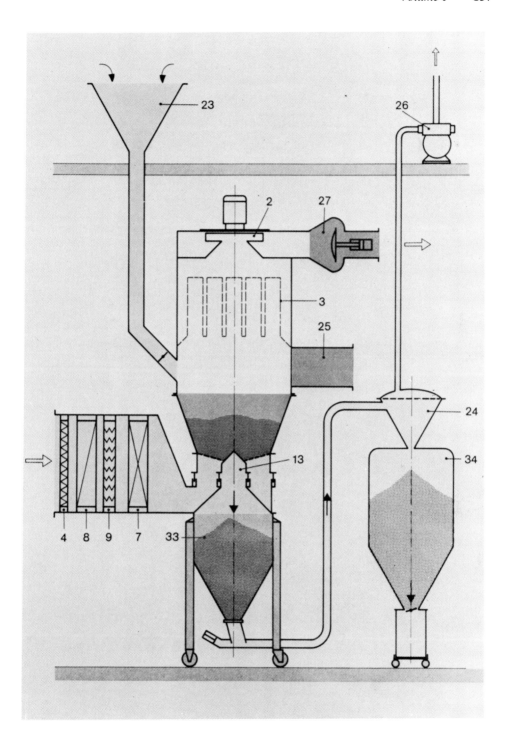

FIGURE 5.   Fluid-bed plant with automatic feeding and discharge. Key as in Figure 4. (From Aeromatic Tech. Brochure No. 78.1.M.30, Aeromatic Inc., Towaco, N. J., 1978. With permission.)

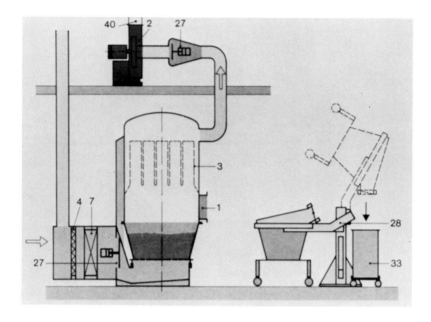

FIGURE 6.   Fluid-bed unit with tilting and discharging device. Key as in Figure 4. (From Aeromatic Tech. Brochure No. 78.1.M.30, Aeromatic Inc., Towaco, N.J., 1978. With permission.)

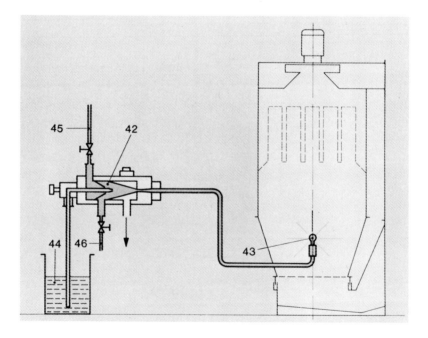

FIGURE 7.   Clean-in-place system. Key as in Figure 4. (From Aeromatic Tech. Brochure No. 78.1.M.30, Aeromatic Inc., Towaco, N.J., 1978. With permission.)

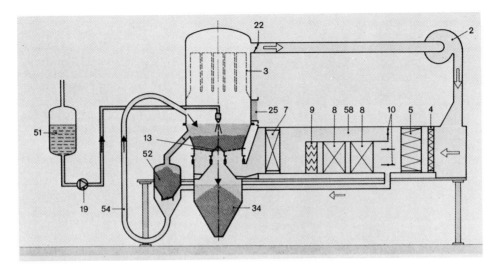

FIGURE 8. Fluid-bed plant with gas circuit within a closed system. Key as in Figure 4. (From Aeromatic Tech. Brochure No. 78.1.M.30, Aeromatic Inc., Towaco, N.J., 1978. With permission.)

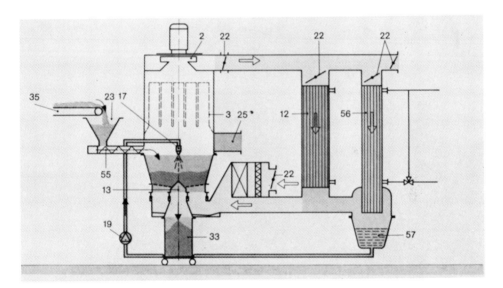

FIGURE 9. Fluid-bed plant with closed gas circuit and integrated product feeding and discharging. Key as in Figure 4. (From Aeromatic Tech. Brochure No. 78.1.M.30, Aeromatic Inc., Towaco, N.J., 1978. With permission.)

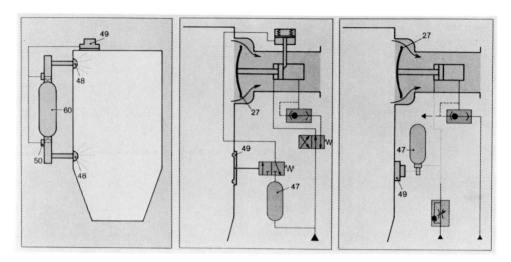

FIGURE 10.   Safety devices: (a) explosion suppression system, (b) quickly locking valve biased by compressed air, and (c) detonating cap-actuated quickly locking valve. Key as in Figure 4. (From Aeromatic Tech. Brochure No. 78.1.M.30, Aeromatic Inc., Towaco, N.J., 1978. With permission.)

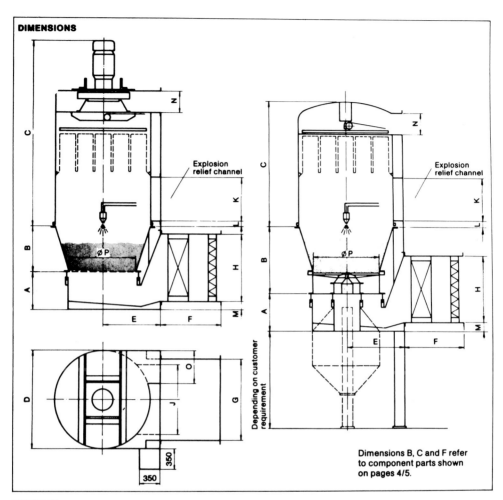

DIMENSIONS

Explosion relief channel

Explosion relief channel

Depending on customer requirement

Dimensions B, C and F refer to component parts shown on pages 4/5.

Size	A	B					C								D	E
		221	222	223	224	225	121	122	123	124	125	126	127	128		
3	256	500	759	759	759	500	1756	1330	1398	1595	1272	1330	1756	1330	700	480
4	321	500	772	772	772	500	1851	1330	1443	1726	1364	1330	1851	1330	900	590
5	411	500	789	789	789	500	2062	1330	1513	2094	1573	1330	2062	1330	1100	670
6	476	550	834	834	834	550	2159	1330	1588	2427	1824	1330	2159	1330	1300	820
7	526	607	917	917	917	607	2624	1815	1815	2648	1983	1815	2624	1815	1500	980
8	566	637	962	962	962	637	2787	1903	2154	3013	2322	1903	2787	1903	1700	1080
9	645	637	955	955	955	637	3336	2260	2551	3240	2415	2260	3336	2260	2000	1230
10	706	657	996	996	996	657	3661	2525	2874	3412	2554	2525	3661	2525	2300	1550

Size	F							G	H	J	K	L	M	N	O	P
	421	422	423	424	425	426	427									
3	620	1475	3200	780	1320	1450	900	495	500	379	450	47	100	135	200	382
4	620	1475	3200	820	1320	1500	900	647	500	567	460	47	100	180	300	565
5	800	1660	3400	1115	1500	1810	1180	786	728	785	450	47	100	250	475	773
6	800	1660	3400	1115	1500	1810	1180	984	728	1003	420	47	100	250	475	991
7	800	1660	3900	1115	1500	1810	1320	1251	1028	1229	510	92	100	250	725	1212
8	800	1660	3900	1365	1500	2090	1380	1412	1128	1410	550	267	150	450	850	1392
9	800	1660	4200	1415	1500	2140	1480	1715	1332	1716	580	328	150	600	1000	1687
10	800	1660	4400	1415	1500	2140	1480	2076	1332	1968	680	363	200	600	1250	1939

FIGURE 11. Dimensions of Aeromatic fluid bed granulator/dryers. (From Aeromatic Tech. Brochure No. 78.1.M.30, Aeromatic Inc., Towaco, N.J., 1978. With permission.)

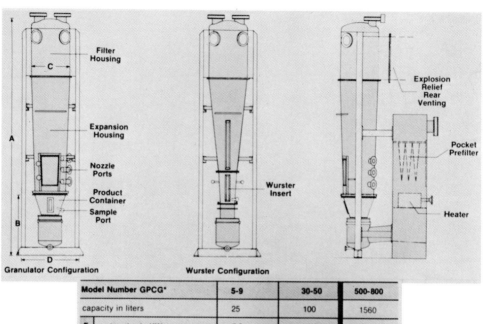

Model Number GPCG*		5-9	30-50	500-800
capacity in liters		25	100	1560
F	motor size in KW	5.5	18.5	55
A	air capacity in cfm	440	1760	7040
N	static pressure inches of water	40	40	40
heating capacity in BTU's per hour		64,000	240,000	1,008,000
DIMENSIONS In MM	A	2700	3500	8400
	B	800	1000	2400
	C	650	720	2600
	D	760	900	3250

FIGURE 12.   Technical specifications and dimensions of the Wurster particle coater/granulator. (From Glatt Tech. Bull. GPCG 8204, Glatt Air Techniques, Ramsey, N.J., 1982. With permission.)

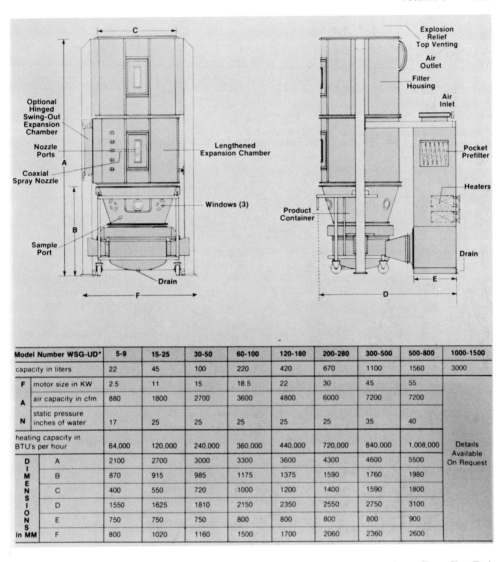

Model Number WSG-UD*		5-9	15-25	30-50	60-100	120-180	200-280	300-500	500-800	1000-1500
capacity in liters		22	45	100	220	420	670	1100	1560	3000
F	motor size in KW	2.5	11	15	18.5	22	30	45	55	
A	air capacity in cfm	880	1800	2700	3600	4800	6000	7200	7200	
N	static pressure inches of water	17	25	25	25	25	25	35	40	
heating capacity in BTU's per hour		64,000	120,000	240,000	360,000	440,000	720,000	840,000	1,008,000	Details Available On Request
D I M E N S I O N S in MM	A	2100	2700	3000	3300	3600	4300	4600	5500	
	B	870	915	985	1175	1375	1590	1760	1980	
	C	400	550	720	1000	1200	1400	1590	1800	
	D	1550	1625	1810	2150	2350	2550	2750	3100	
	E	750	750	750	800	800	800	800	900	
	F	800	1020	1160	1500	1700	2060	2360	2600	

FIGURE 13. Technical specification and dimensions of a Glatt fluid bed granulator/dryer. (From Glatt Tech. Bull. WSG-UD8103, Glatt Air Techniques, Ramsey, N.J., 1981. With permission.)

Table 6
COATING OF PELLETS WITH EUDRAGIT® ACRYLIC RESINS

Active ingredient	Dosage (mg)	Drug content (%)	%dry lacquer substance	% excipients	Eudragit® type	Release rate (time [hr]/% release)			
Amobarbital (sedative)	30	30	6	9	E/RL 1:1	0.5/25	2/50	4.5/75	7/85
Chlorpheniramine maleate (antihistaminic)	2	6	10	10	RL/RS/S 2:2:1	1/25	2/40	4/65	6/90
	2	6	10	2	ED	1/8	2/36	4/64	6/78
Potassium chloride (K-supplement)	500	90	4	4	ED	1/50	2/90		
Lithium carbonate (antidepressive)	300	55	15	10	RL/L 2:1	1/20	2/55	3/75	4/90
PPA HCl (sympathomimetic)	50	30	8	12	RS/S 5:3	1/20	2/30	4/50	6/75
Trifluoperazine HCl (tranquilizer)	2	5	8	12	RS	1/35	2/50	4/70	6/85
Vitamin C (vitamin supplement)	150	50	2	3	RL	1/100 (stabilized)			
	150	50	1.2	4.8	ED	1/100 (stabilized)			
Xanthinol nicotinate (vasodilator)	500	75	7	3	ED/RS/RL 6:5:2	1/25	2/40	4/65	6/85

Note: USP-paddle method substituting half the simulated gastric juice after 1 hr with simulated intestinal juice and so on every hour.

From Lehmann, K. and Dreher, D., *Int. J. Pharm. Technol. Prod. Manuf.*, 2(4), 31, 1981. With permission.

Table 7
COATING OF CRYSTALS WITH EUDRAGIT® ACRYLIC RESINS

Active ingredient	Particle size (mm)	% dry lacquer substance applied	Eudragit® type	% release rate[a] (time [hr]/% release)			
ASA (antiarthritic)	0.3—0.8	2—6	L/S 1:1	1/15	2/27	4/51	6/78 (2%)[b]
	0.3—1.5	10	L	1/1	2/3	3/81	4/92
	0.3—1.25	10	LD	1/0.5	2/0.9	3/73	4/92
Chinidine sulfate (antiarrhythmic)	0.1—0.3	10—20	E	Insulation			
Isoniazid (tuberculostatic)	0.1—0.3	20	RS	1/45	2/65	4/88	6/98
Lithium citrate (antidepressive)	0.25—1.0	25	RS/ED	1/14	2/30	4/59	6/77
Methaqualone (hypnotic)	0.3—0.8	5	RL/RS 1:1	0.5/35	1/50	2/75	3/85[c]
Nitrofurantoin (urinary tract antibacterial)	0.08—0.2	10	L	Enteric coated			
Papaverin (antispasmodic)	0.05—0.3	14	ED	1/27	2/43	4/63;6/78	
Paracetamol (antipyretic)	0.3—0.8	7—10	ED	1/18	2/38	4/61	6/77
Piracetam (antiemitic)	0.1—0.3	7—10	L/E/ED	Insulation			
Potassium chloride (K-supplement)	0.1—0.5	12	ED	1/70	2/95		
Propicillin (antibiotic)	0.1	10	E	Insulation			
Propyphenazone (antipyretic)	0.2—0.5	10	E	Insulation			
Vitamin B$_1$-complex	<0.1	1.3	L	O$_2$-stabilization			
Vitamin C	0.05—1.25	20	L	2/21 (partially enteric coated)			

[a] See legend to Table 6 for release-rate determination method.
[b] See legend to Table 8 for dissolution method.
[c] Dissolution studied in simulated gastric juice only.

From Lehmann, K. and Dreher, D., *Int. J. Pharm. Technol. Prod. Manuf.*, 2(4), 31, 1981. With permission.

Table 8
COATING OF ASA (BAYER 0.7- TO 0.8-MM DIAMETER) WITH EUDRAGIT® L IN ISOPROPANOL

Coating system	Batch size (Kg)	% coating (dry lacquer substance)	Release rate (%)			
			Gastric juice		Intestinal juice	
			1 hr	2 hr	3 hr	4hr
Uniglatt 1	1	8	0.5	1.2	76	90
		10	0.5	0.6	81	92
		12	0.4	0.6	57	70
		14	0.3	0.5	37	55
WSG 5	6	6	3.8	5.5	84	92
		8	2.2	3.3	88	93
		10	2.3	3.7	78	86
WSG 30	50	8	1.0	5.5	84	92
		10	2.1	4.1	85	91
		12	2.2	3.5	87	94

Note: The release rates were determined using the USP-paddle method with 5 g of ASA in 1000 mℓ USP-simulated gastric fluid (without enzymes) stirred at 50 rpm. After 2 hr the particles were filtered off and resuspended in 1000 mℓ USP-simulated intestinal fluid (without enzymes). By pH-stat titration, the pH was kept constant at 7.5 with N-NaOH. Samples of 10 mℓ were hydrolyzed with N-NaOH for 30 min at 100°C at pH 12.0, brought to pH 3.5 with N-HCl, and salicyclic acid estimated colorimetrically with $Fe(NO_3)_3$ at 529 nm.

From Lehmann, K. and Dreher, D., *Int. J. Pharm. Technol. Prod. Manuf.*, 2(4), 31, 1981. With permission.

Table 9
ASA COATED WITH EUDRAGIT® L-30 D

Coating system	Batch size (kg)	% coating (dry lacquer substance)	Release rate (%)				Particle diameter of raw material (mm)
			1 hr	2 hr	3 hr	4 hr	
Uniglatt 1	1	10	1.3	2.3	82	91	1.25
		12	0.7	1.3	83	95	0.6
WSG 30	50	15	1.2	2.0	83	93	0.25—1.25

Note: The release rates were determined using the USP-paddle method with 5 g of ASA in 1000 mℓ USP-simulated gastric fluid (without enzymes) stirred at 50 rpm. After 2 hr the particles were filtered off and resuspended in 1000 mℓ USP-simulated intestinal fluid (without enzymes). By pH-stat titration, the pH was kept constant at 7.5 with N-NaOH. Samples of 10 mℓ were hydrolyzed with N-NaOH for 30 min at 100°C at pH 12.0, brought to pH 3.5 with N-HCl, and salicylic acid estimated colorimetrically with $Fe(NO_3)_3$ at 529 nm.

From Lehmann, K. and Dreher, D., *Int. J. Pharm. Technol. Prod. Manuf.*, 2(4), 31, 1981. With permission.

REFERENCES

1. **Davidson, J. F. and Harrison, D., Eds.,** *Fluidization,* Academic Press, New York, 1971.
2. **Davison, J. F. and Keairns, D. L., Eds.,** Fluidization, in *Proc. 2nd Eng. Found. Conf.,* Cambridge University Press, Cambridge, 1978.
3. **Deasey, P.,** *Microencapsulation and Related Drug Processes,* Marcel Dekker, New York, 1984.
4. **Hall, H. S. and Pondell, R. E.,** The Wurster process, in *Controlled Release Technologies: Methods, Theory, and Applications,* Vol. 2, Kydonieus, A. F., Ed., CRC Press, Boca Raton, Fla., 1980.
5. **Lim, F., Ed.,** *Biomedical Applications of Microencapsulation,* CRC Press, Boca Raton, Fla., 1984.
6. **Thiel, W. J.,** The theory of fluidization and application to the industrial processing of pharmaceutical products, *Int. J. Pharm. Technol. Prod. Manuf.,* 2(4), 5, 1981.
7. **Whitehead, A. B.,** Behavior of fluidized bed systems, *Int. J. Pharm. Technol. Prod. Manuf.,* 2(4), 13, 1981.
8. **Littman, H.,** An overview of flow regimes in fluidized beds, *Pharm. Technol.,* 9(3), 48, 1985.
9. **Scott, M. W., Lieberman, H. A., Rankell, A. S., and Battista, J. V.,** Continuous production of tablet granulations in a fluidized bed. I, *J. Pharm. Sci.,* 53(3), 314, 1964.
10. **Wurster, D. E.,** Means for Applying Coatings to Tablets or Like, U.S. Patent 2,799,241, 1957.
11. **Wurster, D. E.,** Granulating and Coating Process for Uniform Granules, U.S. Patent 3,089,824, 1963.
12. **Wurster, D. E.,** Apparatus for Encapsulation of Discrete Particles, U.S. Patent 3,196,827, 1965.
13. **Wurster, D. E.,** Process for Preparing Agglomerates, U.S. Patent 3,207,824, 1965.
14. **Wurster, D. E.,** Particle Coating Apparatus, U.S. Patent 3,241,520, 1966.
15. **Wurster, D. E.,** Particle Coating Apparatus, U.S. Patent 3,253,944, 1966.
16. **Lindloff, J. A.,** Apparatus for Coating Particles in Fluidized Bed, U.S. Patent 3,117,027, 1964.
17. **Yum, S. I. and Eckenhoff, J. B.,** Development of fluidized-bed spray coating process for axis-symmetrical particles, *Drug Devel. Ind. Pharm.,* 7(1), 27, 1981.
18. **Sarnotsky, A. A.,** Evaporation of solvents from paint films, *J. Paint Technol.,* 41, 692, 1969.
19. **Dean, J. A.,** Vapor pressures of various organic componds, *Lange's Handbook of Chemistry,* 13th ed., McGraw-Hill, New York, 1985, 10.
20. **El-Banna, H. M. and Efimova, L. S.,** The construction and use of factorial design in fluidized bed microencapsulation, *Pharm. Ind.,* 44(6), 641, 1982.
21. **Draper, N. and Smith, H.,** *Applied Regression Analysis,* John Wiley & Sons, New York, 1966.
22. **Myers, R. H.,** *Response Surface Methodology,* Allyn & Bacon, Boston, 1971.
23. **Jones, D.,** Factors to consider in fluid-bed processing, *Pharm. Technol.,* 9(4), 50, 1985.
24. **Mehta, A. M. and Jones, D. M.,** Coated pellets under the microscope, *Pharm. Technol.,* 9(6), 52, 1985.
25. **Pedersen, S. and Moller-Pederson, J.,** Erratic absorption of a slow-release theophylline sprinkle product caused by food, *Pediatrics,* 74, 534, 1984.
26. **Karim, A., Burns, T., Wearley, L., Streicher, J., and Palmer, M.,** Food-induced changes in theophylline absorption from controlled release formulations. I. Substantial increased and decreased absorption with Uniphyll tablets and Theo-Dur Sprinkle®, *Chem. Pharm. Ther.,* 38, 77, 1985.
27. **Amsel, L. P., Hinsvark, O N., Rotenberg, K., and Sheumaker, J. L.,** Recent advance in sustained-release technology using ion-exchange polymers, *Pharm. Technol.,* 8(4), April 1984.
28. **Raghunathan, Y., Amsel, L., Hinsvark, O., and Bryant, W.,** Sustained-release drug delivery system. I. Coated ion-exchange resin system for phenylpropanolamine and other drugs, *J. Pharm. Sci.,* 70(4), 379, 1981.
29. **Raghunathan, Y.,** U.S. Patent 4,221,778, 1980.
30. **Thiel, W. J. and Nguyen, L. T.,** Fluidized bed film coating of an ordered powder mixture to produce microencapsulated ordered units, *J. Pharm. Pharmacol.,* 36, 145, 1984.
31. **Fanger, G. O.,** Microencapsulation processes and applications, in *Proc. Am. Chem. Soc. Symp.,* Vandgaer, J. E., Ed., Plenum Press, New York, 1973.
32. **Goodhart, F. W., Harris, M. R., Murthy, K. S., and Nesbitt, R. U.,** An evaluation of aqueous film-forming dispersions for controlled release, *Pharm. Technol.,* 8(4), April 1984.
33. **Senjkovic, R. and Jalsenjak, I.,** Influence of the atomization time on the properties of ethylcellulose microcapsules of isoniazid prepared by a fluidized bed, *J. Microencapsulation,* 1(3), 241, 1984.
34. **Lehmann, K. and Dreher, D.,** Coating of tablets and particles with acrylic resins by fluid bed technology, *Int. J. Pharm. Technol. Prod. Manuf.,* 2(4), 31, 1981.

Chapter 8

MULTIPLE LAMINATION FOR TRANSDERMAL PATCHES

Dean S. T. Hsieh

TABLE OF CONTENTS

I. INTRODUCTION

The major advantages of rate-controlled transdermal patches include both the convenience of application and removal and the ability to deliver drugs directly to the general circulation, thus bypassing the portal system and the liver. Hence the "first-pass" effect prevalent in orally administered high plasma-clearance drugs is avoided.

In 1980, a transdermal device containing scopolamine was introduced to the market by Alza Corporation (Palo Alto, Calif.). This device provides controlled delivery of scopolamine over a 3-day wearing period. It is effective in the prevention of motion sickness.

In 1982, three different transdermal delivery systems containing nitroglycerin entered the U.S. market. In this form, nitroglycerin has provided effective prophylaxis for both atherosclerotic coronary heart disease and coronary artery spasms. These transdermal devices are successful because they overcome some of the pharmacokinetically undesirable properties of nitroglycerin, e.g., its rapid plasma-elimination half-life and very high plasma clearance. In addition, its high "first-pass" metabolism, which would prevent any pure drug from reaching the general circulation after oral administration, is avoided. Therapeutic use of nitroglycerin had previously been restricted to the treatment of angina attacks as they occurred.[1] All three marketed transdermal devices contain sufficient nitroglycerin to maintain delivery at a more or less constant rate for 24 hr.

Transdermal patches containing nitroglycerin have also been produced abroad. A German product, Deponit®, consists of multiple layers of adhesives with concentration gradients that increase as the layers are further removed from the skin. A Japanese product, Frandel®, is an isosorbide dinitrate patch, wherein the nitroglycerin compound is incorporated into a bubble-forming pressure-sensitive adhesive.

In 1984, a new transdermal patch containing clonadine for antihypertension was approved by the FDA. Other transdermal drug delivery systems are in various stages of development and testing. For example, the Estroderm® patch is designed to release estradiol for the prevention and/or treatment of postmenopausal syndrome.

It is clear that the advent of transdermal patches has been well publicized. The general public as well as the medical community are well aware of the advantages of rate-controlled transdermal administration of drugs. By 1984, sales of nitroglycerin patches had reached $200 million. Because the patent rights on many of the most popular drugs of today have by now expired, incorporation of drugs into transdermal patches seems both promising and profitable. Nevertheless, methods of incorporating drugs into transdermal patches have remained unavailable to potential manufacturers. It is therefore the objective of this chapter to introduce the basic technologies involved in the design and development of transdermal patches.

Most documented fabrication procedures for the design and development of transdermal patches have not yet been disclosed to the public. Researchers in this field are reluctant to do so for two major reasons:

1. The technology involved in the design and development of transdermal patches is very new. Potentially patentable innovations are kept closely guarded.
2. Successful design and development of a useful and convenient transdermal drug delivery system will result in immediate monetary rewards for its designers.

Nevertheless, it is possible to present a general framework through which the design and development of transdermal patches may be discussed (see Figure 1). A transdermal patch is composed of five functional elements (see Table 1):

1. Backing membrane which prevents the drug from migrating outward, instead of in toward the skin

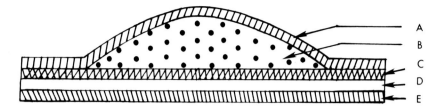

FIGURE 1. A schematic diagram of a transdermal patch: (A) backing membrane, (B) drug reservoir, (C) rate-controlling device, (D) pressure-sensitive adhesive, and (E) release liner.

Table 1
BASIC COMPONENTS OF TRANSDERMAL PATCHES

Components	Types of materials
Backing membrane	Nonwoven, e.g., acrylic blend of polyester and cellulose fibers
	Foam, e.g., 0.0301-in./6-lb density cross-linked polyethylene
	Spunlace, e.g., apertured 100% polyester fabric
	Polyethylene, e.g., 2 mil medium density white polyethylene film
	Paper
	Foil
	Combinations of the above
Drug reservoir[a]	Suspension
	Matrix
	Microsealed
Rate-controlling membrane	Microporous
	Macroporous
	Semipermeable
Pressure-sensitive adhesive[a]	Medical grade rubber
	Natural gum rubber
	Styrene-butadiene
	Polyisobutylene
	Nitrile
	Chloroprene
	Medical-grade acrylate
	Ethyl acrylate
	2-Ethylhexyl acrylate
	Isoocytl acrylate
	Acrylic acid
	Acrylamide
	n-tert-Butyl acrylamide
	Medical-grade silicone
	Polydimethyl siloxanes
	Polydiphenyl siloxanes
	Siloxane blends
Release liner	White Poly Kraft
	Brown Poly Kraft
	Silicone
	Fluorosilicone
	Films

[a]   See following sections for details.

Table 2
MEDICAL USES OF TRANSDERMAL PATCHES

Medical uses	Active drugs	Type of patch (trade name)	Duration (days)
Motion sickness	Scopolamine	Transderm-V®	3
Angina pectoris	Nitroglycerin	Transderm-Nitro®	1
		Nitrodisc®	1
		Nitro-dur®	1
		Deponit®	1
	Isosorbide dinitrate	Frandel®	1
Hypertension	Clonidine	Transderm-Catapres®	7
Postmenopause disease	Estradiol	Estroderm®	Unknown

2.   Drug reservoir, which stores the drug in preparation for rate-controlled release
3.   Rate-controlling device, which limits the amount of drug which is released from the reservoir onto the skin
4.   Pressure-sensitive adhesive, which secures the patch to the surface of the skin
5.   Release liner (peel-off strip), which protects the patch from accidental applications during storing and handling

Together, they comprise a multiple-laminated polymeric skin patch which controls the rate at which a drug dissolves and is transmitted to the surface of the skin.

II. CLASSIFICATION OF TRANSDERMAL PATCHES

A. Classification According to Release Mechanisms
   Within this basic framework there are many variable elements which make classification of transdermal patches possible (see Table 2). For instance, patches may be classified on the basis of their drug release mechanism.[2,3] Membrane-release devices store the drug in a single compartment or reservoir. Matrix-release devices store the drug in a polymeric matrix. Microsealed devices disperse the drug in polyethylene glycol, forming microcompartments or tiny reservoirs. This dispersion is itself dispersed in a matrix of silicone elastomers. In a membrane-release device, the drug reservoir may be composed of pure and solid drug particles or a suspension of solid drug particles in a liquid medium. On one side, the walls of the reservoir are made of an impermeable membrane laminate. On the other side, a nonporous or microporous membrane surrounds the reservoir which controls the rate at which the drug is released from the system (see Figure 2A). The drug migrates through a rate-controlling membrane to the absorption site.
   In developing a matrix-release device, solid drug particles are dispersed in a diffusion-controlling gel or polymer-matrix medium. One side of this matrix remains directly in contact with the skin, while the other side of the matrix is laminated by an impermeable membrane laminate (see Figure 2B). The drug diffuses through the structure to the absorption site.
   In creating a microsealed device, solid drug particles suspended in water-soluble polymers are homogeneously dispersed in a silicone elastomer (see Figure 2C). The elastomer is then cross-linked to form microscopic aliquots of drug solution trapped in the silicone matrix. The matrix may be molded into any shape. Except for the surface which touches the skin, it is enclosed by an impermeable membrane laminate. A nonporous or microporous membrane may provide an additional control on the rate of drug release. Drug release is dependent on the solubility of the drug, the physicochemical properties of the system, and the size and/ or structure of the silicone polymer.

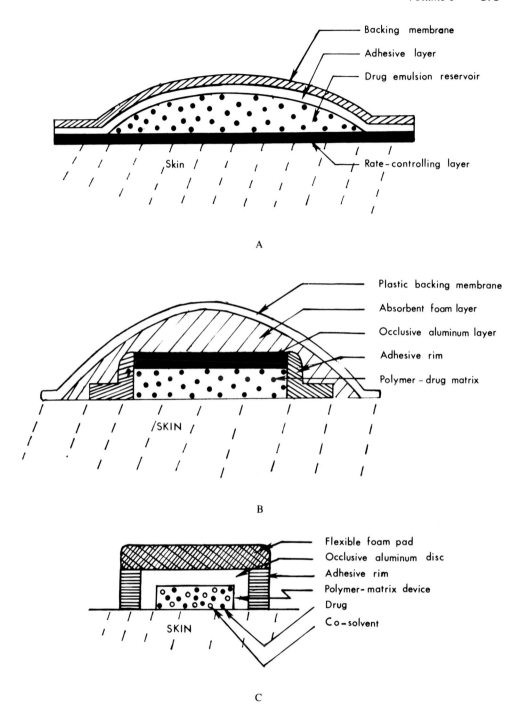

FIGURE 2.   Types of transdermal patches: (A) reservoir type, (B) matrix type, and (C) microsealed type.

All release mechanisms have one common obstacle to overcome: the stratum corneum, or callous layer of the skin. The densely packed, cornified cells of the stratum corneum render it, more than any release mechanism, the rate-limiting barrier to prolonged drug delivery. For this reason, all therapeutic agents in successful transdermal research are biologically active in daily doses of less than 2 mg.[4]

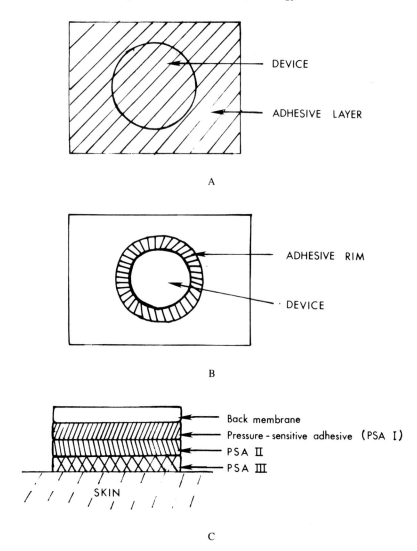

A

B

C

FIGURE 3.   Classification of transdermal patches based upon applying methods for pressure-
sensitive adhesive: (A) crossover type, horizontal view; (B) rim type, horizontal view; and
(C) multiple-layer type, vertical view.

B. Classification According to the Applications of Pressure-Sensitive Adhesives

Another method of classifying transdermal patches is based on their method of utilizing
pressure-sensitive adhesives (see Figure 3). The pressure-sensitive adhesive of a crossover-
type device serves as both an adhesive and a drug delivery vehicle. It encompasses the entire
area between the transdermal device and the skin surface. The drug must pass through it to
reach the skin (see Figure 3A). A rim-type device uses a pressure-sensitive adhesive only
around the periphery of the drug delivery system (see Figure 3B). A membrane or other
mechanism directly contacts the skin. A multiple-layer-type device is held together by the
same pressure-sensitive adhesive which attaches it to the skin (see Figure 3C). The drug is
in solution with the adhesive and is delivered to the skin through diffusion.

C. Classification According to Medical Uses

A third classification system is based on the medical uses of the final device. Currently,
only four medical uses are included in this system: applications for motion sickness, angina

pectoris, hypertension, and postmenopause disease. Motion sickness is treated with scopolamine patches, angina pectoris with nitroglycerin patches, hypertension with clonadine
patches, and postmenopause disease with estradiol patches. All of these drugs have at least
one thing in common: they are therapeutically effective at a daily dosage of approximately
2 mg.[5] Another requirement for transdermal delivery is that the drug be able to permeate
the skin and reach the targeted site; this will be discussed in detail in a later section.

III. DESIGN OF DEVICES

Transdermal delivery devices must be designed with the following criteria in mind:[6]

1. It must not irritate the skin.
2. Its physicochemical characteristics must provide for prompt delivery of drug to the
    skin.
3. It must occlude the skin to insure a one-way flux of the drug.
4. It must exert a bacteriostatic effect to prevent skin flora from proliferating beneath the
    occlusion.

Other factors to be considered concern details such as drug solubility and choice of
pressure-sensitive adhesive. Every decision in the manufacture of a transdermal delivery
system must be made with careful consideration of the consequences, advantages, and
disadvantages of the available alternatives.

A. Membrane Devices

Perhaps the most basic decision to be made in designing a transdermal device is the type
of drug reservoir. Membrane devices have the disadvantage that all of the drug in the system
is stored in one compartment. Were it to burst or leak, the system would be ruined. However,
the advantage of this type of system is that the membrane becomes saturated with the drug
during storage. The release liner prevents it from escaping until it is applied to the skin.
Thus, the skin is saturated upon application of the patch. This priming dose secures the skin
binding sites necessary for continuous, uniform drug delivery.

B. Matrix-Controlled Devices

Both matrix-release and microsealed devices must utilize a different method of achieving
a priming dose, such as mixing the drug in the pressure-sensitive adhesive. However, matrix
devices are safer than the membrane system in that it is unlikely that the drug will escape
from the reservoir. The drug release mechanism of a matrix type device is not simple
diffusion, such as that in membrane release. Rather, the release mechanism in a matrix
device relies on two factors:[7]

1. Drug molecules have a finite solubility in the matrix.
2. Total concentration of drug per unit volume (this includes the undissolved drug) is
    greater than its finite solubility in the matrix.

As drug crystals dissolve into the surrounding medium, the dispersed drug diffuses through
the matrix continuum to the matrix surface. The eluted region of the matrix recedes from
the application site as the process continues.

C. Multiple Layers of Adhesives

Another variation of diffusion-release mechanisms is the multiple-layer transdermal device. In it, the entire dose of drug is combined with the pressure-sensitive adhesive and the

entire patch is composed of layers of the adhesive. The layer closest to the skin has the smallest concentration of drug, and the concentration increases the further the layers are removed from the skin. The drug diffuses in the direction of declining drug density. The design of this transdermal system combines the advantage of priming dose of the membrane release system with the advantage of dispersion of drug of the matrix-release and microsealed system. In principle, multiple layers of adhesives such as in the previously mentioned Deponit® may be classified as matrix-release devices. However, the following advantages of Deponit® render this design more desirable than typical matrix-release devices:

1.   Lamination processes are easier to perform.
2.   Increasing drug concentration gradient compensates for zero-order release.
3.   Release rate is close to drug permeation rate, thus eliminating drug residue on skin.

D. Examples

The relationships among components of a transdermal drug delivery system are easy to discern through the study of specific examples. Transderm-V® is the only type of transdermal scopolamine delivery system on the market.[8] The product is unique in that its pressure-sensitive adhesive contains the drug itself. It provides both adhesion to the skin and a priming dose of scopolamine to saturate skin binding sites for uniform drug delivery. It also includes a steady-state scopolamine reservoir in a polymeric gel of mineral oil and polyisobutylene. The drug is released through a microporous membrane with a pore size which achieves balance between the rate of drug released and the activity level of drug in the reservoir. The drug- and moisture-impermeable backing membrane of Transderm-V® is an aluminized polyester membrane and the peel-off strip is siliconized polyester.

The other drug currently marketed in transdermal patches is nitroglycerin. There are three types of transdermal nitroglycerin.[9] Transderm-Nitro® has a backing membrane of aluminized plastic. Its reservoir contains five times the amount of drug delivered to the body and thus serves as a source of energy for drug diffusion. In it, nitroglycerin is adsorbed on lactose and dispersed in a colloidal suspension of silicone dioxide and silicone medical fluid. The microporous rate-limiting membrane allows 0.5 mg of drug per square centimeter of patch in 24 hr. Although not formulated as such, the pressure-sensitive adhesive of Transderm-Nitro® receives enough of the drug in the patch through diffusion during storage (8%) to function similarly to the pressure-sensitive adhesive of Transderm-V®. Another transdermal nitroglycerin delivery system is Nitro-Dur®. The backing membrane of Nitro-Dur® is composed of four layers, including a polyethylene coverstrip, a moisture-protective absorbent pad, a microporous pressure-sensitive adhesive tape, and a nonpermeable aluminum foil baseplate. The drug reservoir is a three-dimensional polymeric matrix composed of polyvinyl alcohol, lactose, glycerin, water, povidone, and sodium citrate. Nitroglycerin molecules are entrapped in the matrix, gradually dissolve, equilibrate between matrix and liquid phases, and are driven from the matrix as the concentration gradient increases. The third type of nitroglycerin patch is Nitrodisc®. Nitrodisc® utilizes a silicone polymeric matrix as its drug reservoir and release mechanism. Its backing membrane is an aluminum foil disc and it is overlaid by a pressure-sensitive adhesive foam pad, with which it is applied to the skin.

Nitro-Dur® and Transderm-Nitro® deliver nitroglycerin to the skin at the same rate (0.5 mg drug per square centimeter per 24 hr) but Transderm-Nitro® is viewed as more dangerous because the drug reservoir could conceivably rupture. However, the rate of drug delivery of Nitrodisc® is superior to them both (0.625 mg drug per square centimeter per 24 hr). Nitrodisc® also wastes less drug than the other two products. Transderm-Nitro® and Nitro-Dur® contain five and ten times the amount of nitroglycerin which is delivered to the body, respectively. Nitrodisc® contains only three times the amount of delivered drug.

IV. PRESSURE-SENSITIVE ADHESIVES

Pressure-sensitive adhesives are most often used in the manufacture of flexible tapes. An adhesive is spread, sprayed, dipped, or calendered onto a backing element and easily activated through simple manual pressure. Common backing elements are cloth, paper, films, and foils. It is sometimes necessary to treat the backing element with a primer to improve its bond to the adhesive or increase its strength. It may also be necessary to treat the opposite side of the backing element with a release coat to reduce the unwinding tension of a wound tape. The first pressure-sensitive adhesive was designed for medical use in 1845, but applications for pressure-sensitive adhesives now range from protecting surfaces to sealing packages. Over 200 different types of pressure-sensitive adhesives exist today.[10]

Pressure-sensitive adhesives are reducible to five basic components:[11]

1. Elastomer: a fundamental component which may be a high molecular weight synthetic or natural rubber, polyvinyl ether, polyacrylic ester, or block copolymer
2. Tackifier: resins which are added to the elastomer to improve adhesive strength (polyterpene resins, gum resin, rosin ester and other rosin derivatives, oil-soluble phenolic resins, coumarone-indene resins, petroleum hydrocarbine resins)
3. Plasticizer: mineral oil, liquid polybutenes, liquid polyacrylates, or lanolin added to improve low-temperature flexibility, wetting properties, and general cohesion
4. Filler: colors, clays, or other additives which reduce costs, soften, or otherwise affect the final product (zinc oxide, titanium dioxide, aluminum hydrate, calcium carbonate, clay, or pigments)
5. Antioxidant: lends stability to adhesives which are expected to be exposed to high temperatures, which may be rubber, metal dithiocarbamates, or metal-chelating agents

These components are carefully analyzed and combined in a manner geared toward a specific application. Alterable physical properties which may be relevant to specific applications include the following:[12]

1. Tack: the adhesion which forms upon contact between the adhesive and the surface of an object
2. Peel adhesion: the adhesion which is formed between the adhesive and the surface of an object once the bond has reached an equilibrium state
3. Creep resistance: the strength of an adhesive bond

All of these properties are difficult to measure, but a few standard procedures have been employed to judge the effectiveness of one adhesive as compared to another. These evaluation methods are illustrated in Figure 4. Tack may be measured by rolling a stainless steel ball down an inclined plane onto a horizontal surface coated with adhesive. It is expressed in inches of ball travel across the adhesive (see Figure 4A). Peel adhesion may be measured by pressing a strip of adhesive onto a specific surface under specified conditions and peeling it off at a rate of 6 in/min from a 180° angle. It is expressed in the number of pounds of resistance encountered in the peeling process (see Figure 4B). Creep resistance is measured by a method similar to the procedure for measuring peel resistance, but the adhesive is peeled with a two-pound continuous separation at a 90° angle. It is expressed in the number of hours required for the adhesive strip to pull away from the surface (see Figure 4C).

Pressure-sensitive adhesives intended for use in transdermal patches must not only meet physical requirements, but physiological requirements as well. They must not have adverse mechanical, chemical, or allergic effects on the skin. Predictive sensitivity testing serves to prevent the marketing of patches which cause such reactions. Tests for adverse effects are normally divided into four stages:[13]

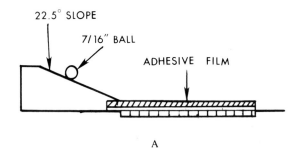

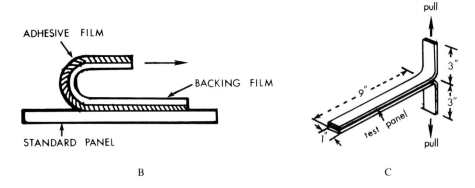

FIGURE 4.   Methods for evaluating pressure-sensitive adhesive: (A) rolling ball tack test, (B) 180° peel adhesion test, and (C) creep resistance test.

1.   Preliminary screening consists of primary irritation tests performed on animals. It includes tests for percutaneous activity and eye irritation.
2.   Preliminary human screening consists of irritancy and sensitization tests on humans.
3.   Actual use of the product under controlled conditions is a necessary prerequisite to marketing.
4.   Limited marketing of the product is desirable to detect possible rare, yet significant, effects.

   Preliminary human screening is the most significant indicator of possible adverse reactions which stem from the use of a given product. Trials of this nature, such as the Draize and Shelanski tests in which 200 subjects are treated with standard patches for 24 hr out of every 48, are carefully devised and diligently carried out. In the Draize test, different sites are used for each application and the testing requires 10 to 24 applications. Subjects then rest for 10 to 14 days before the patch test is challenged at a fresh site to determine sensitization. The Shelanski test is similar to the Draize except that the patches are reapplied to the same site at each of exactly 15 applications, and the rest period lasts for 2 to 3 weeks. Both the Draize test and the Shelanski test determine degree of sensitization through qualitative methods, by judging the degree of irritation at the affected site.
   Others, however, prefer to use more objective, quantitative methods such as ID_{50} and IT_{50} tests.[14] Strong irritants are given an ID_{50} value based on the concentration of substance necessary to produce a discernible reaction in 50% of the test population in 24 hr. Weak irritants are given an IT_{50} value based on the number of days of continuous exposure necessary to identify an irritation reaction in 50% of the test population. In these tests, the experimenter need only decide whether or not a reaction has occurred. The intensity of the reaction is not recorded.

Table 3
CHARACTERISTICS OF FLEXIBLE MATERIALS[19]

Pouch paper	Low cost, rigidity, strength
Glassine paper	Grease resistance, flavor protection
Foil	Moisture and gas protection, good appearance
Cellophane	Stiffness, machinability, transparency
Polyethylene	Low cost, heat sealability
Polypropylene	Moisture barrier, stiffness
Polyvinyl chloride	Grease resistance, heat sealability
Saran	Moisture and gas protection
Rubber hydrochloride	Grease resistance, heat sealability
Polyester	Strength, high- or low-temperature performance
Nylon	Formability for deep draws, toughness

Another significant indicator of the biocompatibility of a given pressure-sensitive adhesive is a photosensitization test.[15,16] Two patches are applied to two symmetrical areas of the body for a 4- to 48-hr period. One patch is removed and the test area is irradiated with 3200 to 4000 Å UV light. The second patch is then removed and that test area is covered with light-impenetrable material. After 48 hr, both sites are examined for erythema, edema, vesicles, and/or bullae. The reactions of the two sites are compared for a final evaluation of photosensitivity.

V. SECONDARY POUCH PACKAGING FOR TRANSDERMAL PRODUCTS

Although the general description of transdermal patches lists five components, a sixth component exists: the package. Transdermal product designers must place great significance on the package because the physical and chemical stability of a transdermal patch is far more delicate than other dosage forms. Sterilized, unit-dose packaging is just beginning to be recognized as labor saving and convenient.[17] It is the preferred method of storing transdermal patches. Because of their delicate nature, transdermal patches must have a package which is strong enough to protect them from distribution environments, but gentle enough to pose no threat to the patch itself.

Both of these requirements are met by form-fill-seal pouches. Advantages of form-fill-seal packaging consist in the fact that the pouches are prelabeled, contain a unit dose, are easy to handle, offer protection from denaturation, and are manufacturable from a wide variety of materials. Limitations stem from the fact that every pharmaceutical use of these pouches requires the approval of the U.S. FDA. The wide variety of materials from which they can be manufactured, however, eases the path of innovation.

The papers, films, and foils used in creating form-fill-seal packages are themselves the subject of current research and the product of innovative technologies. The choice of a metallized film or various laminated constructions is dependent upon both the product and the properties of the final package. Considerations include:[18] (1) moisture content of the product when packaged, (2) desired shelf-life and other storage conditions, (3) mass and texture, (4) number of units in a single package, (5) shipping case quantity, (6) permeability of package by moisture and gases, (7) possible extraction of its plasticizers and stabilizers by the product (or other modifications of package by product), (8) possible absorption of the product by the package, and (9) possible photochemical change of product from exposure to light. Materials are combined according to the desired properties of the final product. In Table 3 the variety of properties which different materials offer is illustrated. However, a combination of two materials does not necessarily yield a combination of their respective properties. Combinations must be tested to determine the extent to which they retain the properties of their components.[20]

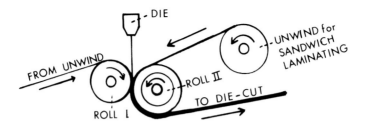

FIGURE 5.   The basic lamination process.

Not only are there many combinations of materials from which to choose, but there is also a wide variety of ways in which to combine them.[21] The basic lamination process consists of two webs joined through an adhesive (see Figure 5). They are passed between rollers to secure the bond. A variation in this process is heat lamination, in which one of the rollers is heated. Another common laminating process is referred to as extrusion coating. Molten plastic (usually low-density polyethylene) is spread over a moving web of paper or film. An extension of the extrusion concept is coextrusion. Both materials of the coextrusion process are molten polymers. They are forced together through a slit die to form a composite film. Increasingly, metallization has been seen as a cost-saving alternative to lamination. Vaporized aluminum bombards film being rolled over a cooled drum in a vacuum. It condenses into a 0.00002-in. layer, and the film, now metallized, is rewound. However, metallized film is most often used in conjunction with some type of lamination.

The materials used in transdermal patches are printed with labels and codes before they are processed into form-fill-seal pouches. Several different types of pouches may be formed once the materials are chosen. Those used for transdermal patches may be one of three types: three-side-seal, four-side-seal, or pillow-pouch packages. Each has a unique manufacturing process.[22]

Three-side-seal pouch packages — These are generally made with horizontal form-fill-seal machines. The machine pulls the web under a downward-pointing isosceles triangle. The edges are then directed up so it is folded in half. A gusset may be folded into the bottom of the folded web by pulling it over a triangular plow. The movement of the web is stopped as pouch sides are heat sealed (leaving flaps at the top) and scissor-like vertical blades separate the pouches. Spring-loaded clamps grab the sides and bring the pouches to an air jet splitter which separates the top flaps. The pouches are opened by a vacuum, duck bill, or wire blade opener. They are filled and reformed with paddles before being heat sealed at the top.

Four-side-seal pouches — These are made by both horizontal and vertical methods. However, for transdermal patch applications, the horizontal systems are preferred because the machinery is less complicated. A web is pulled into the machine and slit lengthwise to form both the top and bottom of the pouch. The patch is placed on the bottom web, and the top web is pulled over it. The sides are sealed with heat and pressure rollers. Rotary sealer bars form the other two seams. Scissor-like knives separate the pouches.

Pillow-pouch packaging — This may also be done horizontally or vertically, but horizontal methods are preferred if the product permits. The web is fed into a former, which makes it into a tube. Tension and tracking are controlled by spring-loaded dancer rollers, guides, clutches, and peelers. A fin seal wheel seals the side edge from the inside of the tube. Mechanical or air tuckers push excess film into the tube to narrow the sealing area. The ends of the pouches are sealed and cut by the same mechanism.

The foremost aim of pouch packaging design is to produce an impermeable, leakproof product. Tests for leakage include the following:[23,24]

1.   The packaged product is submerged in degassed water in a partial vacuum. Air bubbles pinpoint leakage pathways.
2.   The packaged product is submerged in a dye solution in a partial vacuum. The vacuum is released, resulting in a partial vacuum in the package and atmospheric pressure outside of it. Inspection of package contents for dye reveals permeability.
3.   Helium is injected into the package prior to closure. Mass spectrometer detectors determine helium leakage rates and pathways. Permeation rates for plastic films will be directly proportional to the partial pressure differential of the diffusant and leakage rates will be directly proportional to the square of that partial pressure differential.
4.   Dry ice is placed in the package before sealing and carbon dioxide leakage detected by the PAC GUARD 4000 (Modern Controls, Inc.).
5. CO_2 and water vapor permeation rates may be measured by gas chromatography or infrared detection.
6.   Oxygen leakage rates may be determined by gas chromatography or coulometric detection.

VI. CONTINUOUS PROCESS FOR MULTIPLE LAMINATION

The manufacture of transdermal patches is a continuous and fully automated process. Layer by layer, materials are formed by an assembly line into a completed product[25] (see Figure 6). Production of a membrane-release device begins with a web of impermeable backing material unwinding to form an assembly line. The drug suspension is fed from a temperature-controlled mixer into an injection pump. A multihead injector sets the drug on the backing membrane as it moves down the line. A prelaminated web composed of an impermeable release liner, a layer of pressure-sensitive adhesive, and a rate-controlling membrane is unwound on top of the drug-laden backing membrane. Molding machinery forms the layers into the final product. The final product is fed directly into packaging machinery and the process is complete.

Production of a matrix-release transdermal device begins with a multilaminated web which consists of a rim-type pressure-sensitive adhesive and a release liner. The drug matrix is preformulated with polyvinyl acetate (PVA) glycerol, water, and a water-soluble polymer. It is molded into cylinders which are fed into a slicing machine. As the multilaminated web is unwound onto an assembly line consisting of an impermeable laminate, the slicing machine slices and drops a specified thickness of the cylindrical matrix onto the web. An impermeable backing membrane unwinds over the device to complete the process. The device is then fed into packaging machinery.

Production of a microsealed device begins with an impermeable backing membrane and an adhesive foam layer. They are unwound together to form an assembly line. In a separate process, the drug suspension is mixed with an elastomer, a viscosity modifier, and a cross-linking agent. This mixture is fed through an extruder to a multihead injector. The multihead injector places the preformulated microreservoir of drug onto the backing membrane side of the combination. The release liner is unwound over these materials and they are molded together to form the final product. The product is fed into packaging machinery to complete the process.

Production of a device which utilizes multiple layers of adhesives begins with the impermeable backing membrane. It is unwound from a web to form an assembly line. A liquid pressure-sensitive adhesive with the highest concentration of drug is applied and the adhesive solvent is evaporated in a conventional drying tunnel.[26] The process is repeated as the coated backing membrane is coated with a pressure-sensitive adhesive of decreasing drug concentrations. Once the required number of coats in decreasing concentration gradients has been applied, the web is die-cut in spherical shapes and fed directly into pouch-packaging machinery.

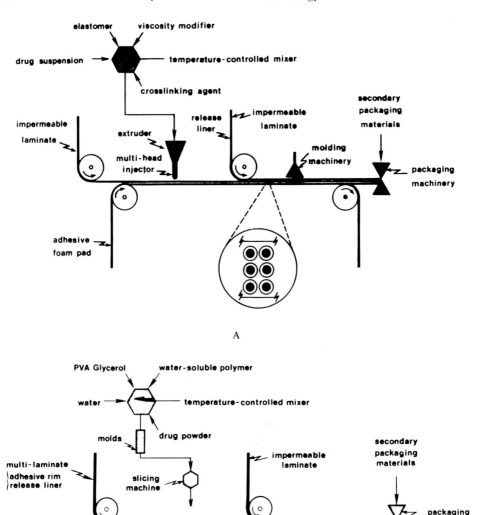

FIGURE 6.   Continuous process for multiple lamination: (A) process for manufacturing reservoir-type transdermal patches, (B) process for manufacturing matrix-type transdermal patches, (C) process for manufacturing microsealed-type transdermal patches, (D) process for manufacturing transdermal patches made of multiple layers of adhesives. (Courtesy of Dr. Yie Chien, College of Pharmacy, The State University, Rutgers, New Brunswick, N.J.)

VII. SKIN ADJUVANTS FOR TRANSDERMAL PATCHES

The most basic difficulty which researchers in transdermal technology must face is that the natural function of human skin is to retain body fluids and to protect the body from hazardous chemicals. It thus presents a natural barrier to transdermal drug administration. Specifically, the stratum corneum (the outermost layer of the skin) is the rate-limiting barrier.

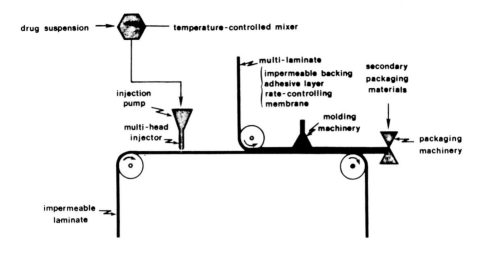

FIGURE 6C.

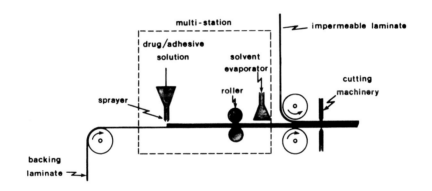

FIGURE 6D.

It consists of approximately 15 layers of cornified cells, each of which is surrounded by a proteinaceous envelope.[27] The main ingredient of these cells is keratin. Their shape is hexagonal or pentagonal. They are tightly bound by lipid-like membranes, rendering it possible to enhance permeation through the use of lipophilic drug carriers.

Once through the stratum corneum, drug molecules must partition into and diffuse through the viable epidermis. The diffusional properties of this layer are similar to an aqueous protein gel.[28] From the viable epidermis, drug molecules enter a microcirculatory network which delivers them to systemic circulation.

Substances which pave the way for drug diffusion through the stratum corneum are called skin adjuvants[29-68] (see Table 4). A few methods of enhancing skin permeability have been documented, but the mechanisms of permeation are yet unclear. For instance, treatment of the epidermis with surfactants is known to improve the penetration of polar molecules while a two-component system of polar solvent and lipid is required to achieve the same penetration for nonpolar molecules. This leads researchers to believe that there are distinct, yet undiscovered, pathways through the stratum corneum. As another example, the difference in pH between the dermal and epidermal layers of the skin has been identified as a potential transport mechanism for anionic drug molecules.

A particularly successful experimental skin adjuvant is 1-dodecylazacycloheptane-2-one, or Azone®. Azone® is a colorless, relatively odorless, and minimally irritating substance

Table 4
PERCUTANEOUS ABSORPTION ENHANCERS

Enhancer	Test system	Drug	Factor X	Ref.
Urea	Human skin	Hydrocortisone		29
	Human skin	Triamcinolone acetonide		29
Dimethylsulfoxide (DMSO)	Mouse skin	Ethidium bromide		30
	Mouse skin	Dichlorotriazinyl (procion) dyes		30
Calcium thioglycolate	Rat skin	Theophylline	40	31
DMSO	Rat skin	Theophylline	9	31
Polyoxyethylene lauryl ether (B1-gex)	Rat skin	Theophylline	9	31
Sorbitan trioleate	Rabbit in vivo	Flufenamic acid		32
Polyoxyl (8) stearate	Rabbit in vivo	Flufenamic acid		32
Polyoxyethylene (2) oleyl ether	Rabbit in vivo	Flufenamic acid		32
DMSO	Rabbit in vivo	Flufenamic acid		32
Sorbitan monopalmitate + DMSO	Rabbit in vivo	Flufenamic acid		32
Sorbitan trioleate + DMSO	Rabbit in vivo	Flufenamic acid		32
Polyoxyl (8) stearate + DMSO	Rabbit in vivo	Flufenamic acid		32
Polyoxyethylene 20	Rabbit in vivo	Flufenamic acid		32
Cetyl ether + DMSO	Rabbit in vivo	Flufenamic acid		
Polyoxyethylene (2) oleyl ether + DMSO	Rabbit in vivo	Flufenamic acid		32
DMSO	Rat skin	Salicylic acid		33
2-Pyrrolidone	Human skin	Methanol		34
	Human skin	Octanol		34
	Human skin	Caffeine		34
Dimethylformamide	Human skin	Methanol		34
	Human skin	Octanol		34
	Human skin	Caffeine	15	34
N,N-Diethyl-m-toluamide	Mouse skin	Hydrocortisone		35
	Human skin	Hydrocortisone		35
1-Dodecylazacycloheptane-2-one (Azone®)	Mouse skin	Clindamycin phosphate		36
	Mouse skin	Erythromycin base		36
	Mouse skin	Fusidate sodium		36
	Mouse skin	Fluorouracil		36
	Mouse skin	Desonide		36
	Mouse skin	Amcinonide		36
	Mouse skin	Triamcinolone acetonide		36
Benzyl nicotinate	Rat in vivo	Dexamethasone	1.5	37
DMSO	Human skin	Sarin		38
Dimethylacetamide	Human skin	Sarin		38
Dimethylformamide	Human skin	Sarin		38
Formamide	Human skin	Sarin		38
Dioxane	Human skin	Sarin		38
Methylorthoformate	Human skin	Sarin		38
Urea	Guinea pig skin	5-Fluorouracil		39
Propylene glycol + decanol	Human skin	Salicylic acid		40
Propylene glycol + hexanol	Human skin	Salicylic acid		40
Propylene glycol + fatty acid	Human skin	Salicylic acid		40
Propylene glycol + oleyl alcohol	Human skin	Salicylic acid		40
Propylene glycol + oleyl methyl sulfoxide	Human skin	Salicylic acid		40
Hexane + ethylene glycol monomethylether	Rat in vivo	Diazepam		41
3,1,1,3,3-Tetramethyl urea + hexamethyldisiloxane	Rat skin	Diazepam		42
2-Cyclohexyl-1,1,-dimethyl-ethanol	Human in vivo	Lidocaine HCl		43

Table 4 (continued)
PERCUTANEOUS ABSORPTION ENHANCERS

Enhancer	Test system	Drug	Factor X	Ref.
Ricinoleyl alcohol ethoxylated partial glyceride of C6-12 saturate fatty acid	Guinea pig in vivo	Bromovinylarabino-furanosyluracil		44
1-Dodecylazacycloheptan-2-one (Azone®)	Skin	Propatyl nitrate		45
1-Butylazacyclopentan-2-one	Skin	Propatyl nitrate		45
N-Methyl-2-pyrrolidone	Human in vivo	Betamethasone 17-benzoate		46
	Rabbits	Mefenamic acid		68
1-Dodecylazacycloheptane-2-one (Azone®)	Skin, mucous membranes	Wide variety of drugs		47
	Mouse skin	Clindamycin		48
	Mouse skin	Erythromycin		48
	Mouse skin	Fusidic acid		48
	Mouse skin	Triamcinolone acetonide		48
	Mouse skin	Desonide		48
	Mouse skin	Amcinonide		48
	Mouse skin	Desoxymetazone		48
	Mouse skin	Fluocinolone acetonide		48
	Mouse skin	Anthracene		48
	Mouse skin	8-Bromo cyclic adenylic acid		48
	Mouse skin	5-Fluorouracil		48
Urea	Rat in vivo	Indomethacin		49
Eucalyptol	Mouse skin	Procaine		50
N,N-Diethyl-m-toluamide	Mouse in vivo	Hydrocortisone		51
1-Dodecylazacycloheptane-2-one (Azone®)	Mouse skin	Clindamycin phosphate		52
	Mouse skin	Erythromycin		52
	Mouse skin	Fusidate Na		52
	Mouse skin	Fluorouracil		52
	Mouse skin	Desonide		52
	Mouse skin	Amcinonide		52
	Mouse skin	Triamcinolone acetonide		52
Methyl octylsulfoxide	Rabbits in vivo	Isosorbide dinitrate		53
Alkyl sulfones	Rabbits in vivo	Isosorbide dinitrate		53
Alkyl acetamides	Rabbits in vivo	Isosorbide dinitrate		53
N,N-Diethyl-m-toluamide	Mouse skin	Hydrocortisone		54
Sucrose monooleate	Mouse skin	Tetracycline HCl		55
Diethyl sebacate	Rat skin	Salicylic acid		56
Dimethyl sulfoxide	Rat skin	Salicylic acid		56
Trichloroethanol	Mouse skin	Ethynylestradiol-3-methyl ether		57
Trifluoroethanol	Mouse skin	Ethynylestradiol-3-methyl ether		57
2-Hydroxyundecyl methyl sulfoxide	Human in vivo	n-Butyryl-scopolamine HBr		58
Decylmethylsulfoxide	Human in vivo	Antiperspirants		59
Poly [2-(methylsulfinyl) etheracrylate]	Skin	Dimethoate		60
	Skin	Paraoxon		60
	Skin	P-Aminohipporic acid		60
Decylmethylsulfoxide	Skin	Hydrocortisone		61
		Methapyrilene-HCl		61
DMSO	Skin	Glucocorticoids		62
Propylene glycol	Human skin	Estradiol		63
	Human skin	Metronidazol		63
Ethylene glycol	Human skin	Estradiol		63
	Human skin	Metronidazol		63

Table 4 (continued)
PERCUTANEOUS ABSORPTION ENHANCERS

Enhancer	Test system	Drug	Factor X	Ref.
Hexadecanol	Human skin	Estradiol		64
Octadecanol	Human skin	Estradiol		64
Hexamethyldisiloxane	Rabbit skin	Diazepam		65
		Clorinidine base		65
Decamethyltetrasiloxane	Rabbit skin	Indomethacin		65
Cyclic dimethylsiloxanes	Rabbit skin	Indomethacin		65
Polyethylene glycol 400	Mouse skin	Estradiol		66
1-Dodecylazacycloheptane-2-one (Azone®)	Mouse skin	Triamcinolone acetonide		67
	Silicone membrane	Triamcinolone		67
Dialkyl sulfoxides	Human in vivo	Scopolamine HBr		61
Nicotinamide	Rabbit	Mefenamic acid		68

which has been demonstrated to enhance the percutaneous absorption of antibiotics, glucocorticoids, and other dermatologic drugs. Pretreatment of the skin with Azone® has been shown to further enhance its adjuvant properties. Although the mechanism by which Azone® performs this task is unknown, it is apparent that its absorption by the skin carries with it a "reservoir effect", in which the drug is retained in the stratum corneum even when drug administration ceases. This is thought to be the effect of skin hydration, but it is not known why azone in particular is able to capitalize on this effect.

Because skin adjuvants such as Azone® increase the number of drugs which may permeate the stratum corneum in therapeutically active doses, their incorporation into transdermal patches is simply a matter of time. In the future, transdermal patch design will involve decisions such as whether the skin adjuvant should be released from the pressure-sensitive adhesive or should be an individual component of the device. The answer may be that devices with crossover-type or multiple-layer pressure-sensitive adhesives should contain the adjuvant within the adhesive, while devices with rim-type pressure-sensitive adhesives should utilize a separate reservoir. The use of adjuvants in transdermal patch technology will not only widen the range of drugs available for transdermal delivery, but will have important consequences for future patch designs.

VIII. PROPOSED TOXICOLOGICAL TESTS FOR TRANSDERMAL PATCHES

The previously described toxicity test programs for pressure-sensitive adhesives are of general application. However, a very rigid primary toxicity test program must be developed specifically for transdermal drug delivery systems. It must include a primary irritation assay, a sensitization test, a dermal toxicity study, and a bacteriological study.

The primary irritation assay involves six dorsal-clipped rabbits.[69] Each is treated with four patches: a control patch on an abraded area, a control patch on an unabraded area, a medicated patch on an abraded area, and a medicated patch on an unabraded area. The patches are removed after 24 hr and each of the 24 areas is measured for degree of erythema and edema at 24 and 72 hr postapplication (none: 0, severe: 4). The primary irritation index is calculated by taking the mean values of the six rabbits for abraded and unabraded areas. Both control and medicated patches are reduced to four values of erythema and four for edema. The primary irritation index per animal is determined by summing the eight values for each rabbit and then dividing by four. If it is greater than five, positive irritation has occurred.

The sensitization test (Buehler Test) utilizes 60 dorsal-dilapidated guinea pigs.[70] Three groups of 20 are treated with a control patch, a medicated patch, and a 2,4-dinitrochlorobenzene patch as a positive control. The patches are applied for 6-hr periods on day 0, 7, and 14 and for 24 hr on day 28. Treated areas are assigned degrees of erythema (none: 0, severe: 3). The percentage of animals with scores greater than or equal to 1 is determined. Sensitization is measured for each type of patch depending on the determined sensitization rate. There are five grades of sensitization: 0 to 8%, 9 to 28%, 29 to 64%, 65 to 80%, and 81 to 100%.

The dermal toxicity study utilized 3 groups of 20 dorsal-dilapidated rats (10 male, 10 female). One group is not treated. One is treated with a nonmedicated patch. The third is treated with a medicated patch. Patches are replaced once a week for 4 weeks. Parameters to be monitored include body weight, food consumption, clinical chemistries, urinalysis, and, in the fourth week, organ weights, hematology, and histopathological exams.

Bacteriological studies must be done under the supervision of clinicians. The patches must be incubated in a culture plate. Both the nature and quantity of bacterial colonies which develop must be evaluated. The results of such tests may indicate one of two phenomena: (1) the patches have been contaminated during the manufacturing or packaging process or (2) the final product itself provides an environment conducive to bacterial growth.

Secondary toxicological tests for transdermal patches include the Draize, Shelanski, IT_{50}, and photosensitization tests described above. The same tests which determine the toxicity of pressure-sensitive adhesives may be used to determine the toxicity of actual patches. These tests are sufficient to determine toxicity because these patches are developed using previously tested, FDA-approved drugs.

IX. CONCLUSIONS

Multiple lamination for transdermal patches employs a congregation of existing technologies from many fields: pharmaceutical science, chemical engineering, adhesive technology, packaging technology, and medical science. Combined, these fields converge on a new drug administration route: the skin. When the skin is seen as a route of drug administration, its very size opens a wide range of options for delivery devices. However, the science of percutaneous drug delivery is still extremely underdeveloped. Much research must be completed on skin physiology before the design of transdermal devices may be evaluated with specificity and certainty.

Therefore, the technologies explored in this chapter are admittedly insufficient to describe this rapidly growing field. Nonetheless, systematic study of these methods will refine them to the point of precise quality control. The technologies covered in this chapter include (1) transdermal device design, (2) applications of pressure-sensitive adhesives for transdermal patches, (3) packaging options for transdermal products, (4) manufacturing processes for existing transdermal products, (5) skin adjuvant technologies, and (6) toxicological tests for transdermal patches. The chapter thus provides a general framework for understanding multiple lamination processes for transdermal patches. Future research in this field will enrich this framework and further its development. In a recent advance, a new class of chemical entity was discovered by the author at Conrex Pharmaceutical Corporation: skin permeation enhancers. Due to patent protection policies, however, their chemical structure cannot be disclosed.

REFERENCES

1. **Goldman, P., Ed.,** *Proc. Symp. Transdermal Delivery of Cardiovascular Drugs,* C. V. Mosby, St. Louis, 1984; as cited in *Am. Heart J.,* 108(1), 195, 1984.
2. **Karim, A.,** Transdermal absorption: a unique opportunity for constant delivery of nitroglycerin, *Drug Dev. Ind. Pharm.,* 9(4), 671, 1983.
3. **Chien, Y.,** Logics of transdermal controlled drug administration, *Drug Dev. Ind. Pharm.,* 9, 579, 1983.
4. **Shaw, J.,** Pharmacokinetics of Nitroglycerin and Clonidine Delivered by the Transdermal Route, *Amer. Heart J.,* 108(1), 217, 1984.
5. **Shaw, J.,** Pharmacokinetics of Nitroglycerin and Clonidine Delivered by the Transdermal Route, *Amer. Heart J.,* 108(1), 222, 1984.
6. **Parikh, N. H., Babar, A., and Plakogiannis, F. M.,** Transdermal therapeutic systems, II, *Pharm. Acta Helv.,* 60(2), 34, 1985.
7. **Keshary, P. R.,** Mechanism of Transdermal Controlled Nitroglycerin Administration — Control of Skin Permeation Rate and Optimization, Doctoral thesis, Rutgers University, New Brunswick, N.J., 1984, 59.
8. **Parikh, N. H., Babar, A., and Plakogiannis, F. M.,** Transdermal Therapeutic Systems, Part 2, *Pharm. Acta Helv.,* 60(2), 35, 1985.
9. **Parikh, N. H., Babar, A., and Plakogiannis, F. M.,** Transdermal therapeutic systems, II, *Pharm. Acta Helv.,* 60(2), 36, 1985.
10. **Bemmels, C. W.,** Pressure sensitive tapes and labels, in *Handbook of Adhesives, 2nd ed.,* Skeist, I., Ed., Van Nostrand-Reinhold, New York, 1977, 724.
11. **Barth, B. P.,** Phenolic resin adhesives, in *Handbook of Adhesives, 2nd ed.,* Skeist, I., Ed., Van Nostrand-Reinhold, New York, 1977, 414.
12. **Herman, B. S.,** *Adhesives — Recent Developments,* Noyes Data Corporation, Park Ridge, N.J., 1976, 3.
13. **Fisher, A. A.,** Sensitivity testing, in *Cosmetics, Science, and Technology,* Vol. 3, 2nd ed., Balsam, M. S. and Sagarin, E., Eds., John Wiley & Sons, New York, 1974, 295.
14. **Fisher, A. A.,** Sensitivity testing, in *Cosmetics, Science, and Technology,* Vol. 3, 2nd ed., Balsam, M. S. and Sagarin, E., Eds., John Wiley & Sons, New York, 1974, 303.
15. **Epstein, S.,** The photopatch test, *Ann. Allergy* 22(1), 1, 1964.
16. **Fisher, A. A.,** Sensitivity testing, in *Cosmetics, Science, and Technology,* Vol. 3, 2nd ed., Balsam, M. S. and Sagarin, E., Eds., John Wiley & Sons, New York, 1974, 290.
17. **Kelsey, R. J.,** *Packaging in Today's Society,* Saint Regis Paper, Bangor, Maine, 1978.
18. **Hanlon, J. F.,** *Handbook of Package Engineering,* 2nd ed., McGraw-Hill, New York, 1984, 13.
19. **Hanlon, J. F.,** *Handbook of Package Engineering,* 2nd ed., McGraw-Hill, New York, 1984, 11.
20. **Hanlon, J. F.,** *Handbook of Package Engineering,* 2nd ed., McGraw-Hill, New York, 1984, 16.
21. **Hanlon, J. F.,** *Handbook of Package Engineering,* 2nd ed., McGraw-Hill, New York, 1984, 8.
22. **Davis, G. C.,** *Packaging Machinery Operations: Form-Fill-Sealing,* Packaging Machinery Manufacturers Institute (PMMI), Washington, D.C., 1982.
23. **Amini, M. A.,** Testing permeation and leakage rates of pharmaceutical containers, *Pharm. Technol.,* p. 39, December 1981.
24. **Amini, M. A. and Morrow, D. R.,** Leakage and permeation: theory and practical applications, *Package Dev. Syst.,* p. 20, May/June, 1979.
25. **Chien, Y. W.,** The use of biocompatible polymers in rate-controlled drug delivery systems, *Pharm. Technol.,* p. 50, May 1985.
26. **Butler, J. P.,** Laminations, *Packag. Encycl. Yearb.,* 30(4), 72, 1985.
27. **Kligman, A. M.,** Skin permeability: dermatological aspects of transdermal drug delivery, *Am. Heart J.,* 108(1), 200, 1984.
28. **Guy, R. H. and Hadgraft, J.,** Transdermal drug delivery: the ground rules are emerging, *Pharm. Int.,* p. 112, May 1985.
29. **Wohlrab, B.,** The influence of urea on the penetration kinetics of topically applied corticosteroids, *Acta Derm. Venereol.,* 64(3), 233, 1984.
30. **Shackleford, J. M., Yielding, K. L., and Scherff, A. H.,** Observations on epidermal exsorption in mice following injections of procion dyes and ethidium bromide and topically applied dimethylsulfoxide, *J. Invest. Dermatol.,* 82(6), 629, 1984.
31. **Kushida, K., Masaki, K., Matsumura, M., Ohshima, T., Yoshikawa, H., Takada, K., and Muranishi, S.,** Application of calcium thioglycolate to improve transdermal delivery of theophylline in rats, *Chem. Pharm. Bull.,* 32(1), 268, 1984.
32. **Hwang, C. C. and Danti, A. G.,** Percutaneous absorption of flufenamic acid in rabbits: effect of dimethyl sulfoxide and various nonionic surface active agents, *J. Pharm. Sci.,* 72(8), 857, 1983.
33. **Rossi, G. A., Zamboni, A. M., and Persichella, M.,** Interference of extraneous solutes in the iontophoresis of salicylic acid across the skin, *Boll. Soc. Ital. Biol. Sper.,* 59(6), 806, 1983.

34. **Southwell, D. and Barry, B. W.,** Penetration enhancers for human skin mode of action of 2-pyrrolidone and dimethylformamide on partition and diffusion of model compounds water, n-alcohols, and caffeine, *J. Invest. Dermatol.,* 80(6), 507, 1983.

35. **Windheuser, J. J., Haslam, J. L., Caldwell, L., and Shaffer, R. D.,** The use of N,N-diethyl-M-toluamide to enhance dermal and transdermal delivery of drugs, *J. Pharm. Sci.,* 71(11), 1211, 1982.

36. **Stoughton, R. B.,** Enhanced percutaneous penetration with 1-dodecylazacycloheptan-2-one, *Arch. Dermatol.,* 118(7), 474, 1982.

37. **Kassem, M. A. and Schulte, K. K.,** Effect of benzyl nicotinate on the percutaneous absorption of dexamethasone in the rat, *Eur. J. Drug Metab. Pharmacokinet.,* 5(1), 25, 1980.

38. **Matheson, L. E., Wurster, D. E., and Ostrenga, J. A.,** Sarin transport across excised human skin, II. Effect of solvent pretreatment on permeability, *J. Pharm. Sci.,* 68(11), 1410, 1979.

39. **Wohlrab, W.,** Effect of urea on the mechanism of percutaneous permeation, *Dermatologica,* 159(6), 441, 1979.

40. **Cooper, E. R.,** Increased skin permeability for lipophilic molecules, *J. Pharm. Sci.,* 73(8), 1153, 1984.

41. Nitto Electric Industrial Co. Ltd., Transdermal Pharmaceuticals Containing Drug Absorption Enhancers, Jpn. Kokai Tokkyo Koho, JP 59/95212 A2, 1984.

42. Nitto Electric Industrial Col., Ltd., Transdermal Pharmaceuticals Containing Siloxanes and Urea Derivatives, Jpn. Kokai Tokkyo Koho, JP 59/53408 A2, 1984.

43. **Sipos, T.,** Alcoholic Potentiating Agent for Skin Penetration of Topical Drugs, Australian Patent AU 534455 B2, 1984.

44. **Kreiner, C. F. and Loebering, H. G.,** Dermatological Preparation with Improved Bioavailability, German Patent, DE 3233638 Al, 1984.

45. Nitto Electric Industrial Co., Ltd., Transdermal Tapes Containing Accelerators of Drug Transport in the Skin, Jpn. Kokai Tokkyo Koho, JP 58/208216 A2 1983.

46. **Barry, B. W., Southwell, D., and Woodford, R.,** Optimization of bioavailability of topical steroids: penetration enhancers under occlusion, *J. Invest. Dermatol.,* 82(1), 49, 1984.

47. **Rajadhyaksha, V. J.,** Penetration Enhancers for Transdermal Drug Delivery of Systemic Agents, U.S. Patent 4405616 A, 1983.

48. **Stoughton, R. B. and McClure, W. D.,** Azone: a new non-toxic enhancer of cutaneous penetration, *Drug Dev. Ind. Pharm.,* 9(4), 725, 1983.

49. **Daikyo, Y. and Kogyo, K. K.,** Transdermal Formulations Containing Urea, Jpn. Kokai Tokkyo Koho, JP 58/52216 A2, 1983.

50. **Zupan, J. A.,** Use of Eucalyptol for Enhancing Skin Permeation of Bioaffecting Agents, U.S. Patent 280967, 1981.

51. **Windheuser, J. J., Haslam, J. L., Caldwell, L., and Shaffer, R. D.,** The use of N,N-diethyl-M-toluamide to enhance dermal and transdermal delivery of Drugs, *J. Pharm. Sci.,* 71(11), 1212, 1982.

52. **Stoughton, R. B.,** Enhanced percutaneous penetration with 1-dodecylazacycloheptane-2-one, *Arch. Dermatol.,* 118(7), 475, 1982.

53. Nitto Electric Industrial Co., Ltd., Transdermal Films Increasing Drug Transport into the Skin, Jpn. Kokai Tokkyo Koho, JP 57/59805 A2, 1982.

54. **Windheuser, J. J., Haslam, J. L., Caldwell, L. J.,** Composition of Matter for Topical Application Comprising a Bioaffecting Agent, U.S. Patent 127,881, 1980.

55. **Smith, D. E.,** Dermatological Compositions Containing a Sugar Ester and a Sulfoxide or Phosphine Oxide, U.S. Patent 4046886, 1977.

56. **Sanchez de Rivera Vazquez, A., Rodriguez Perez, A., and Izquierdo San Jose, M.,** Study of a penetrating agent in percutaneous absorption. I. Comparative study using dimethylsulfoxide, *Cienc. Ind. Farm.,* 9(5), 133, 1977.

57. **Higuchi, T.,** Sympathomimetic Topical and Percutaneous Administration with Halogenated Promoters, U.S. Patent 3968245, 1976.

58. **MacMillan, F. S. and Lyness, W. I.,** Compositions for Topical Applications to Animal Tissue and Method of Enhancing Penetration by Them, U.S. Patent 3953599, 1976.

59. **MacMillan, F. S. and Lyness, W. I.,** Compositions for Topical Application of Animal Tissue, U.S. Patent 3903256, 1975.

60. **Hofmann, V., Aingsdorf, H., and Muacevic, G.,** Pharmacologically active polymers, 8-poly 2-(methylsulfinyl) ethylacrylate and their mediating effect on the resorption of pharmaceuticals through the skin, *Makromol. Chem.,* 176, p. 1929, 1975.

61. **MacMillan, F. S. and Lyness, W. I.,** Compositions for Topical Application to Animal Tissue and Enhancing Penetration Thereof, U.S. Patent 3839566, 1974.

62. **Herschler, R. J.,** Enhancing Tissue Penetration of Physiologically Active Steroidal Agents with Dimethylsulfoxide, U.S. Patent 3711606, 1973.

63. **Mollgaard, B. and Hoelgaard, A.,** Vehicle effect on topical drug delivery. I. Influence of glycols and drug concentration on skin transport, *Acta Pharm.,* 20(6), 433, 1983.

64. **Mollgaard, B. and Hoelgaard, A.,** Vehicle effect on topical drug delivery. II. Concurrent skin transport of drugs and vehicles components, *Acta Pharm. Suec.,* 20, 443, 1983.
65. **Sato, S., Kobayashi, I., Nishiu, Y., and Nitto Denki Kogyo, K. K.,** Base Compositions and Pharmaceutical Compositions for External Application, Japanese Patent SHO 58-12102, 1983.
66. **Valia, K. H. and Chien, Y. W.,** Long-term skin permeation kinetics of estradiol. I. Effect of drug solubilizer — polyethylene glycol 400, *Drug Dev. Ind. Pharm.,* 10(7), 951, 1984.
67. **Chow, D. S.-L., Kaka, I., and Wang, T. I.,** Concentration-dependent enhancement of 1-dodecylaza-cycloheptan-2-one on the percutaneous penetration kinetics of triamcinolone acetonide, *J. Pharm. Sci.,* 73(12), 1794, 1984.
68. **Naito, S. I., Nakamori, S., Awataguchi, M., Nakajima, T., and Tominaga, H.,** Observation on and pharmacokinetic discussion of percutaneous absorption of mefenamic acid, *Int. J. Pharm.,* 24, 127, 1985.
69. **Draize, J. H.,** Dermal toxicity, in *Appraisal of the Safety of Chemicals in Food, Drugs, and Cosmetics,* Association of Food and Drug Officials of the U.S., Austin, Tex., 1959.
70. **Buehler, E. V.,** Experimental skin sensitization in the guinea pig and man, *Arch. Dermatol.,* 91, 171, 1965.

INDEX